Reproductive Ethics

Lisa Campo-Engelstein • Paul Burcher
Editors

Reproductive Ethics

New Challenges and Conversations

Springer

Editors
Lisa Campo-Engelstein
Alden March Bioethics Institute
and Department of Obstetrics
& Gynecology
Albany Medical College
Albany, NY, USA

Paul Burcher
Alden March Bioethics Institute
and Department of Obstetrics
& Gynecology
Albany Medical College
Albany, NY, USA

ISBN 978-3-319-52629-4 ISBN 978-3-319-52630-0 (eBook)
DOI 10.1007/978-3-319-52630-0

Library of Congress Control Number: 2017934882

Printed on acid-free paper

This Springer imprint is published by Springer Nature
The registered company is Springer International Publishing AG
The registered company address is: Gewerbestrasse 11, 6330 Cham, Switzerland

Contents

About the Contributors

Kavita Shah Arora received a B.S. from the Pennsylvania State University, an M.D. from Jefferson Medical College, and a Master's in Bioethics from the University of Pennsylvania. She is Board certified in Obstetrics and Gynecology. She serves as the Director of Quality for the Department of Obstetrics and Gynecology at MetroHealth Medical Center and an Assistant Professor of Reproductive Biology and Bioethics at Case Western Reserve University.

Department of Obstetrics and Gynecology, MetroHealth Medical Center, Case Western Reserve University, Cleveland, OH, USA

Paul Burcher is an Associate Professor of Bioethics and Obstetrics and Gynecology at Albany Medical College. He is the Program Director for the residency in Obstetrics and Gynecology, and the Associate Director of the Alden March Bioethics Institute. He is a practicing generalist obstetrician/gynecologist with over 21 years in practice in both private and academic settings. Since 2011 he has worked in a joint appointment between the departments of Obstetrics and Gynecology and Bioethics at Albany Medical College. His publications have been on the doctor-patient relationship, physician empathy, and more recently on ethical issues in clinical obstetrics, including cesarean delivery on maternal request (CDMR), birth plans, and home birth. He is currently involved in a research project studying cesarean section regret.

Alden March Bioethics Institute, Department of Obstetrics and Gynecology, Albany Medical College, Albany, NY, USA

Lisa Campo-Engelstein is an Associate Professor at the Alden March Bioethics Institute and Department of Obstetrics and Gynecology at Albany Medical College in Albany, New York. She earned a Ph.D. from Michigan State University in philosophy with a focus in bioethics and feminist theory. She completed a postdoctoral fellowship with the Oncofertility Consortium at Northwestern University Feinberg

School of Medicine. Her main research area is reproductive ethics, especially contraception, childbirth, fertility preservation, and assisted reproductive technologies.

Alden March Bioethics Institute, Department of Obstetrics and Gynecology, Albany Medical College, Albany, NY, USA

Steph deNormand graduated from St. Lawrence University with a B.S. in Biology and is expected to complete an M.A. in Women's, Gender, and Sexuality Studies from University at Albany in May 2017. During their time at the University at Albany they are the acting Coordinator of the WGS Teaching Collective, teaching feminist pedagogy and overseeing the instruction of Introduction to Feminisms. Their main academic, advocacy, and activism concentration focuses on gender identity and expression, trans identities, and LGBTQ and diversity education.

Women's, Gender, and Sexuality Studies Department, University at Albany, Albany, NY, USA

Alesha Doan is an Associate Professor in the School of Public Affairs and Administration and the Political Science Department at the University of Kansas. Professor Doan's interdisciplinary research is guided by her broader interests in public policy, organizations, and gender/social equity. Her publications include articles in a variety of journals as well as two books, *Opposition and Intimidation: The Abortion Wars and Strategies of Political Harassment* (University of Michigan Press, 2007) and *The Politics of Virginity: Abstinence in Sex Education* (Praeger, 2008). Professor Doan is currently working on a book, *For Their Own Protection: Abortion Regret and the Attack on Women's Reproductive Freedom,* with Professor Shoshanna Ehrlich. She is the lead researcher on a 3-year grant (2016–2019) from the Department of Health and Human Services, Office on Women's Health, to develop and implement policies and prevention programs addressing the issue of sexual assault on college campuses.

School of Public Affairs and Administration and Political Science, University of Kansas, Lawrence, KS, USA

J. Shoshanna Ehrlich is a Professor in the Women's and Gender Studies Department at the University of Massachusetts Boston. Her scholarship focuses on the sexual and reproductive rights of young women and abortion law. Legal in its focus, her work is also highly interdisciplinary in its commitment to situating the law in an historical and social context. In addition to the publication of many law review articles, her books include *Who Decides? The Abortion Rights of Minors* (Praeger, 2006) and *Regulating Desire: From the Virtuous Maiden to the Purity Princess* (SUNY, 2014), as well as a family law textbook. Professor Ehrlich is currently working on a book, *For Their Own Protection: Abortion Regret and the Attack on Women's Reproductive Freedom,* with Professor Alesha Doan. She also works with a variety of advocacy organizations committed to securing the reproductive and sexual rights of young women.

Women's and Gender Studies, University of Massachusetts Boston, Boston, MA, USA

Jazmine Gabriel is a practicing genetic counselor, currently specializing in cancer genetics. She is also an Assistant Professor of Bioethics at Albany Medical College's Alden March Bioethics Institute. She earned a Ph.D. in philosophy from the University of Oregon and an M.S. in genetic counseling from Boston University. Her current research interests include home birth midwives' perceptions of and experiences with genetic testing and genetic counselors, genetic counselors' perspectives on interpreting direct-to-consumer genetic testing, conflict of interest in genetic counseling, and the concept of information in direct-to-consumer genetic testing.

Genetic Counseling Program, Boston University School of Medicine, Boston, MA, USA

Karey A. Harwood is an Associate Professor of Religious Studies in the Department of Philosophy and Religious Studies at North Carolina State University. She is also the Coordinator of NC State's Women's and Gender Studies program. She received her Ph.D. from the Graduate Division of Religion at Emory University where she studied religious ethics and the sociology of religion and morality and was awarded a Woodrow Wilson Dissertation Grant in Women's Health. The focus of her research has been in biomedical ethics and reproductive technologies. Her book, *The Infertility Treadmill: Feminist Ethics, Personal Choice, and the Use of Reproductive Technologies* (UNC Press), examines both the social context of reproductive technologies and the individual experience of infertility, including the tendency of reproductive technologies to function as a consumeristic rite of passage in coming to terms with infertility. Her recent work has focused on egg freezing.

Department of Philosophy and Religious Studies, North Carolina State University, Raleigh, NC, USA

Hermine Hayes-Klein is a lawyer and the Founder of Human Rights in Childbirth, an international nonprofit that connects grassroots activists with legal advocacy to advance women's rights in maternity care. Hermine received her B.A. and J.D. degrees from the University of Chicago. She spent 6 years practicing family law, LGBT rights advocacy, and art law in New York and Connecticut before moving to the Netherlands, where her children were born at home with a Dutch midwife in 2007 and 2010. Hermine taught international law at the Hague University from 2008 to 2012, where she was also Director of the Bynkershoek Institute Research Center for Reproductive Rights. Hermine lives in Portland, Oregon, with her family.

Law Offices of Hermine Hayes-Klein, LLC, Portland, OR, USA

Human Rights in Childbirth, The Hague, The Netherlands

Claire Horner is a Clinical Ethics Fellow at the Alden March Bioethics Institute at Albany Medical College. She earned a J.D. and M.A. in Bioethics from the University of Pittsburgh, and prior to that earned a B.A. in Philosophy and Theology from Franciscan University of Steubenville. She has experience speaking both regionally and internationally on issues in bioethics, including legal and ethical issues in foregoing life-sustaining treatment, informed consent, and reproductive technologies. Her research is currently focused on reproductive ethics and law, particularly issues in embryo donation and gestational surrogacy.

Alden March Bioethics Institute, Albany Medical College, Albany, NY, USA

Allan J. Jacobs received a B.A. in Psychology from Cornell University, an M.D. from the University of Southern California, and a J.D. from St. John's University. He is Board certified in Obstetrics and Gynecology and in Gynecologic Oncology. He has served as Chair and Residency Program Director at two large hospitals in New York City. He currently is Director of Gynecologic Oncology at Coney Island Hospital, Brooklyn, New York, Professor of Obstetrics and Gynecology and Associate Faculty in Bioethics, Stony Brook University, Stony Brook, New York, and Adjunct Professor at New York Medical College.

Gynecologic Oncology, Coney Island Hospital, Brooklyn, NY, USA

Department of Obstetrics and Gynecology, Coney Island Hospital, Brooklyn, NY, USA

Department of Obstetrics and Gynecology, Stony Brook University, Stony Brook, NY, USA

Kirsten M. Kringle-Baer received her Ph.D. in philosophy from the University of Minnesota with a dissertation on epistemic reasons for not always pursuing knowledge. She is particularly interested in situations that arise in the context of seeking or providing health care and in the ethics and epistemology of nursing practice. She has taught at the University of Minnesota, the University of Saint Catherine, and the Kansas City University of Medicine and Biosciences.

Cincinnati, OH, USA

Susan B. Levin is Roe/Straut Professor in the Humanities and a Professor of Philosophy at Smith College. The focus of her research is Greek philosophy and bioethics, which she is keen to bring together. Her publications in Greek philosophy include two books and numerous articles. In the final chapter of *Plato's Rivalry with Medicine: A Struggle and Its Dissolution* (Oxford, 2014), Levin argues that drawing on Plato, particularly the *Laws*, can enrich current bioethical discussions of the doctor-patient tie, medical professionalism, and medicine's relationship to society. Levin's current work focuses on bioethical controversy regarding enhancement. In her explorations of the debate over vigorous enhancement as framed and pursued thus far, she often finds it illuminating to draw on Greek philosophy, especially Plato and Aristotle. In addition to publishing articles on enhancement, Levin is currently writing a book on the topic.

Department of Philosophy, Smith College, Northampton, MA, USA

Tara A. Lynch is currently a first year fellow in Maternal Fetal Medicine at the University of Rochester Medical Center. She completed her undergraduate degree in biochemistry at Siena College. She then attended medical school at Albany Medical College and remained at Albany Medical Center for her residency in Obstetrics and Gynecology. She currently involved in research on preterm birth specifically pertaining to periviability and placental infections.

Department of Obstetrics & Gynecology, University of Rochester Medical Center, Rochester, NY, USA

Reproductive Ethics: New Challenges and Conversations Introduction

Lisa Campo-Engelstein and Paul Burcher

When we first envisioned organizing a conference solely on issues in reproductive ethics we were met by some skepticism by friends and colleagues. Were there really enough compelling topics and scholarship in this area to sustain interest for a 2-day conference? Our view, and the perspective confirmed by this book, is that reproductive ethics is a field that has many questions, which we have only started to explore to the extent that they deserve. There are two reasons this is true. One is that some "hot button" issues have received all the attention—abortion, in particular, comes to mind here. The other is that unlike many well-litigated areas of bioethics, such as end-of-life care, emerging technologies in reproductive ethics are changing the ethical playing field faster than the bioethics literature can keep pace. As for whether reproductive ethics can sustain interest we can only note that it is hard to imagine a subject that humans spend more time and energy upon than sex and reproduction. It is the only area of bioethics that can be called sexy without invoking an implausible metaphor.

We have organized this book into sections that share a common theme, and the first theme explored is "Prenatal Genetic Testing and Enhancement." A philosopher who is also currently training to be a genetic counselor wrote the first chapter. Jazmine Gabriel addresses the tension between maternal autonomy and the disability community, and the way that prenatal genetic counselors seek to navigate this tension without disrespecting either side. Her perspective is that the non-directive counseling techniques taught to genetic counselors do not escape the critique of disability advocates because women and couples are still expected to decide whether to continue a pregnancy based upon genetic testing that reveals only one

L. Campo-Engelstein (✉) • P. Burcher
Alden March Bioethics Institute, Department of Obstetrics and Gynecology,
Albany Medical College, Albany, NY, USA
e-mail: CampoeL@mail.amc.edu; BurcheP@mail.amc.edu

L. Campo-Engelstein, P. Burcher (eds.), *Reproductive Ethics*,
DOI 10.1007/978-3-319-52630-0_1

aspect of the fetus—a metonymy of sorts where a part comes to stand for the whole. While some aspects of this dilemma between individual choice and respect for the disability community may seem insoluble, she suggests that the lens of solidarity points to new perspectives that may lessen the tension between these two apparently opposing values.

Susan Levin's chapter, "Enhancing Future Children: How it Might Happen, Whether it Should," addresses proponents of "Procreative Beneficence" and uses Aristotelian philosophy to critique its privileging of cognitive ability over a more balanced notion of human faculties. She argues that human flourishing cannot be enhanced solely by boosting human rationality, and that Aristotle recognized that human "thought and character" represent a "harmony between … (the) … rational and non-rational." Reading this chapter we are reminded that reaching back 2500 years one can still find philosophical descriptions of our humanity that have relevance to current debates regarding emerging genetic technologies.

The final chapter in this section, "The Decision to Know: Pregnancy and Epistemic Harm" by Kirsten M. Kringle-Baer, examines the epistemic harms of knowledge that can be garnered through prenatal ultrasound, screening, and testing during pregnancy. The author recounts her pregnancy that was misidentified as high risk, which resulted not only in unnecessary medical interventions and increased stress and anxiety, but also in epistemic harms. In analyzing her pregnancy narrative, Kringle-Baer highlights how the circumstances of both the subject and the object of knowledge can affect whether pursuing knowledge is advisable.

The next section, "Fertility Preservation," focuses on the rapidly emerging and expanding field of fertility preservation. Fertility preservation of an individual's gametes first rose to prominence in the oncology context. Cancer patients of reproductive age were interested in freezing their gametes so they could genetically reproduce later in life; the field of oncofertility was born. Today fertility preservation has expanded beyond cancer patients to include a wide variety of individuals. In this section, the authors focus on fertility preservation for transgender and gender nonconforming individuals and for single women seeking to delay childbearing.

Nonmedical egg freezing has received much attention in the media. In the chapter, "Egg Freezing and the Feminist Quest for Equality in the Workplace," Karey Harwood explores the related topics of nonmedical egg freezing, women's employment, and delayed childbearing. One of the common arguments in favor of nonmedical egg freezing is that it functions as a form of "reproductive affirmative action," allowing women time to focus on their education and careers and still have children using their own eggs. Harwood rejects this argument because its narrative of planning and control does not hold up, especially in light of the health risks associated with egg freezing and its relatively low probability of success. Rather than addressing the larger social issues at play, nonmedical egg shifts the burden onto individual women, which may prevent broader structural changes from occurring in the workplace that could actually engender long-lasting justice.

In the chapter, "Fertility Preservation in Transgender and Gender Nonconforming Youth and Adolescents," Steph deNormand examines the topic of fertility preservation for transgender and gender nonconforming youth and adolescents, a topic that has not received much attention in the reproductive ethics literature. Gender affirming treatments for individuals with gender dysphoria, such as hormonal treatment and surgery, can diminish and even eliminate fertility. Fertility preservation may be an option before beginning gender affirming treatments. However, it raises some ethical issues, especially because these individuals are minors. Drawing upon queer theory, trans theory, and feminist theory, and comparing fertility preservation for transgender and nongender conforming adolescents to fertility preservation for young oncofertility patients, deNormand analyzes the ethical issues involved and makes recommendations for how to handle different types of cases.

The third section focuses on "Childbirth." Hermine Hayes-Klein, a lawyer working in reproductive rights, makes a strong case against the system of maternity care in America, but also lays out a clear vision of how things can be improved. Her arguments are empirical and historical, and her solutions are balanced, in the sense that she recognizes that fully empowering women in their medical choices must also come with some increased responsibility for their now autonomous choices, thus potentially lessening physician liability—particularly when women make choices different than physician recommendations. Her descriptions of her own home births in the Netherlands point to a kind of holistic, even spiritual experience, that women have difficulty accessing in this country. It is hard to read her work without thinking that we can, or must, do better.

Tara Lynch, a fellow in maternal-fetal medicine, and Paul Burcher, an obstetrician/gynecologist and bioethicist (and co-editor of this book) write on the ethical challenges of periviable birth—births in the range of extreme prematurity. The question they address is whether a cesarean section can be justified by maternal decision-making that is not grounded in medical beneficence, but rather in an expanded sense of beneficence that includes psychosocial, cultural, and other factors. The example discussed is whether it is reasonable to perform an elective cesarean section if it allows the birth of a living baby, even if the death of the child from prematurity is ultimately a certainty. Arguing from other analogous cases, they conclude that an expanded notion of beneficence can justify some elective cesarean sections in the periviable period.

The field of reproductive ethics includes many "hot button" issues, some of which do not fit into the previous sections. In this final section, "Abortion, Surrogacy, and Circumcision," the authors address three important "hot button" issues. The first chapter in this section, "'Teaching Morality by Teaching Science:' Religiosity and Abortion Regret" by Alesha Doan and J. Shoshanna Ehrlich, deals with the longstanding controversial topic of abortion. Specifically, the authors focus on how the concept of abortion regret is being used to limit access to abortion including the 2007 *Carhart v. Gonzales* decision, in which United States Supreme Court drew upon abortion regret to support upholding the federal ban on late-term abortions. After examining 180 affidavits from women who experienced abortion regret as well as interviewing 25 antiabortion activists, the authors conclude that the concept of abortion regret exists as intertwined with and inherently a part of these women's religious beliefs. While they acknowledge the importance of these women's experiences, they assert that their religious perspectives should not be treated as universal for all women and that abortion regret should not be upheld as a legitimate empirical claim.

The second chapter in this section, "My Child, Your Womb, Our Contract: The Failure of Contract Law to Protect Parties in Gestational Surrogacy" by Claire Horner, tackles another contentious topic, commercial gestational surrogacy. In the United States, contract law is used for commercial surrogacy. Yet, as Horner points out, contract law is poorly suited for reproductive arrangements like surrogacy: Important components of surrogacy contracts are unenforceable, the standard of performance is not clear, and damages for a breach of contract are inadequate. Moreover, contract law may complicate the clinical management of a surrogacy pregnancy and may fail to honor the reproductive autonomy of all the individuals involved. Due to these shortcomings, she concludes that contract law is not a suitable framework for governing commercial surrogacy arrangements.

The third chapter in this section, "Vulvar Nick and Metzitzah B'Peh: Punishment or Harm Reduction?" by Allan J. Jacobs and Kavita Shah Arora, addresses the hotly debated topic of genital alteration of infants. They compare two types of genital alteration—vulvar nick and metzitzah b'peh (direct oral suction of blood from the penile head that is used in some Jewish ritual circumcisions)—and conclude that while neither may be ethically justifiable, a punitive criminal approach to them is not defensible from both a utilitarian perspective and a retributive justice perspective. Instead, they assert a harm reduction approach as the best way to respond to vulvar nick and metzitzah b'peh.

We believe that the chapters offered in this compilation represent a fair representation of the breadth and depth of discussions within the field of reproductive ethics, yet we also recognize that there are still many questions and controversies not addressed here. There is work to be done, and the accelerating technologies of reproduction and genetics make this task both urgent and exciting.

Part I
Genetic Testing and Enhancement

Zooming Out: Solidarity in the Moral Imagination of Genetic Counseling

Jazmine Gabriel

Introduction

Prenatal genetic testing allows potential parents to screen for and diagnose an ever-increasing number of conditions. However, because there is no treatment for the majority of currently detectable conditions, the information opens the door to selective termination, which some in the disability community consider morally problematic. Genetic counselors, and the other genetics professionals who offer these tests, have a professional obligation to help each client[1] navigate the implications of genetic information and to make a decision, which, while legally considered a personal and private decision, inevitably spills over into the political realm.

In this paper, I begin by giving an overview of the nature of the tension between the disability community and the genetic counseling profession. However, my goal is not to side with one group or the other. Instead, I would like take a step back and consider the lens through which the tension between the two communities has been viewed, namely, in terms of the principle of autonomy. While I do not intend to question the importance of reproductive freedom, I would like to ask whether the language of choice and freedom fully captures what is going on here. In brief, I argue that the principle of autonomy as the primary framework for understanding both the problem and potential responses to the problem, is insufficient, and that looking to the relational underpinnings of autonomy may shed greater light on

[1] Historically, genetic counselors have used the term "client" rather than "patient" because the psychosocial aspects of the profession derive from Rogerian client-centered therapy and because many founding figures in the profession were Ph.D.s rather than M.D.s, and wanted to make clear that they did not claim to be treating sick patients. Today, genetic counselors tend to use the terms interchangeably. I use the term "client" out of respect for the original intention.

J. Gabriel (✉)
Genetic Counseling Program, Boston University School of Medicine, Boston, MA, USA
e-mail: gabriel2@bu.edu; jazgabriel@gmail.com

L. Campo-Engelstein, P. Burcher (eds.), *Reproductive Ethics*,
DOI 10.1007/978-3-319-52630-0_2

previous responses and future possibilities. Shifting the theoretical lens will not resolve the tension between the two communities; but perhaps looking at the problem from different angles will broaden our ability to imagine currently unforeseen alternatives.

In particular, and following the 2015 article by Bruce Jennings and Angus Dawson entitled "Solidarity and the Moral Imagination of Bioethics," my aim is to consider if or how the concept of solidarity[2] might enrich thinking about the tension between the genetic counseling profession and the disability community. I consider the role of the concept of solidarity in terms of what the authors refer to as the "calculus of consent" and the "moral imagination," which I will interpret broadly as tools for understanding the past and imagining the future.

Why solidarity? Jennings and Dawson note that "[i]n contemporary neoliberal societies … ideological currents are promoting a calculus of consent based on interests that separate rather than interests that join" (2015: 32). In controversies where consensus is elusive, where both sides have a legitimate concern, perhaps it is worthwhile to zoom out, rather than remain mired in a yes or no tug of war, and to shift from an exclusive focus on interests that separate to a broader view that contextualizes individual interests and makes apparent the conditions for their possibility.

Certainly state programs in recent history could dissuade anyone from using the language of solidarity; however, promoting the interests of groups over the rights of individuals is not my intention. The aspects of the concept of solidarity that interest me here are the following: Solidarity as the structural context of individual freedom and solidarity as rooted in historical memory. Both aspects situate individuals within a complex network of relationships to other human beings, including those who lived long ago and those who will live in the future. Jennings and Dawson refer to solidarity as a "shaping sensibility" that "informs other normative principles and ideals rather than supplementing or competing with them, a perspective that gives a relational[3] interpretation to regulative ethical principles such as justice, liberty, and beneficence" (2015: 32). Because reproductive freedom is at the heart of debates about the limits of prenatal genetic testing, it is particularly important to view the concept of solidarity as shaping the analysis rather than replacing the principle of autonomy.

[2] See Prainsack and Buyx 2012 for an in depth analysis of the uses of the term solidarity in bioethics literature.

[3] There is an enormous body of literature on ethical theories that critique excessive individualism and prioritize relationships. See, for instance, the ethics of care developed by Carol Gilligan (1982) and Nel Noddings (1984). Many contemporary analyses by phenomenologists, feminists, postmodernists, social-justice theorists, cultural theorists, etc., start from the notion that individual relationships constitute or are prior to the individual. See, for instance, Mackenzie and Stolijar (2000), "Relational Autonomy: Feminist Perspectives of Autonomy, Agency, and the Social Self." See also Kenny et al. (2010) for a summary of the inadequacies of the dominant individualistic approach and a discussion of relational autonomy, relational social justice, and relational solidarity in the context of public health.

The Disability Critique of Prenatal Genetic Testing

The disability[4] critique of prenatal testing centers around two primary claims: "that prenatal genetic testing following by selective abortion is morally problematic, and that it is driven by misinformation" (Parens and Asch 2000: 13). Screening for Down syndrome has been around the longest and exemplifies some of the thornier aspects of the critique. When a pregnant woman is offered diagnostic testing (amniocentesis or chorionic villus sampling), she may use this information to prepare for a child with Down syndrome, but she may also use the information to decide whether or not to pursue termination. Although genetic counselors would like to view themselves as offering the neutral facts, many people feel that the implication in offering the test is that it may be preferable to avoid having a child with this condition (Saxton 2000; Patterson and Satz 2002). Some families who have children with Down syndrome, and some individuals with Down syndrome, feel that testing communicates[5] a fundamental disrespect for the lives of individuals with Down syndrome (Bauer 2005).

Adrienne Asch has characterized the moral wrong of selective abortion as reducing a whole person to a single trait: As in other forms of discrimination when a person is reduced to her sex, color, or sexual orientation, selective abortion communicates that one quality is sufficient to render one person's life less valuable than other lives (Asch 2000). As Martha Saxton (1998) puts it, the message is that "some of us are 'too flawed' in our very DNA to exist" (391). The objection here, called the *expressivist objection*, is that prenatal genetic testing sends a message that it is better not to exist at all than to exist with a disability. Asch argues that the moral wrong lies not in the

---

[4] I follow Adrienne Asch in using the term to include "all health-related departures from species-typical functioning" (Asch 2003: 319, Note 10). See also the definition of disability in the Americans with Disability Act of 1990. While I will use broad terms such as disability community and disability movement, I am aware that groups "are not unified monoliths," as Joseph Stramondo puts it prior to clarifying his intention to use the term "disability movement." He writes, "This term is deliberately broad and meant to encompass the substantial range of sometimes divergent tactics and ideologies deployed by disabled people, but it is not assumed that all disabled people engage in such action, have the same political beliefs, or use identical advocacy approaches. In addition, while the lived reality of disability is an important feature of my argument, the 'disability movement' is not simple shorthand for people who experience life with an anomalous embodiment or medical impairment but instead refers to a particular subset of disabled people who are conscious of their own subordinate social position and engage in political action accordingly" (Stramondo 2016).

[5] The claim that there is a "message" communicated in the offering of or existence of prenatal genetic tests is best understood in the context of the contrast between the medical and social models of disability. If much of the reason why people with disabilities are un-able is rooted in the lack of societal accommodation, then offering prenatal testing conveys a message that society would rather address the challenges associated with disabilities by preventing people with disabilities from being born rather than taking measures to improve institutions and practices that could enhance the lives of people with disabilities. See Asch 2003, "Disability Equality" in *Prenatal Testing. See also* Susan Wendell (1996) who says selective abortion sends the message that "we do not want any more like you" (in *The Rejected Body*).

choice of termination itself, but in the choice to abort *this* particular fetus in cases where the pregnancy is otherwise desired. She calls this the *any/particular* distinction: To terminate because it is not the right time is to say 'I don't want *any* baby right now,' whereas to terminate following prenatal testing, whether because of sex or genetic condition, is to say 'I don't want this particular baby' (Asch 2000; Saxton 2000).

Some take the argument a step further and say that the mere existence and availability of prenatal testing constitutes a judgment of people currently living with the conditions for which there is testing (Madeo et al. 2011; Patterson and Satz 2002). In offering the test, the medical professional communicates that this is a condition whose impact is sufficiently devastating as to warrant prevention altogether. True, clients are not forced to get these tests, or terminate on the basis of results if they do pursue testing, but often the social impact of the existence of the tests is to pressure parents to do both (Munger et al. 2007; Press 2000). For instance, parents who have carried pregnancies with fetuses affected by trisomy 21 or spina bifida report being asked whether they got "the test" (Bauer 2005). Awareness that they will face this question, and that the determination of sympathy or judgment hinges on the answer, inevitably constrains the "choice" facing parents (Jennings 2000; Lippman 2003).

A further dimension to the disability critique of prenatal testing is that by reducing the number of people with genetic conditions, like Down syndrome, we may decrease the social urgency of improving disability accommodations and reducing discrimination (Wasserman and Asch 2006). This perpetuates the medical model of disability, which assumes that the problem is with the particular trait rather than with society's unwillingness to increase accommodations for people with that trait (Saxton 2000; Kaplan 2000).

Messages Sent and Received

The disability critique of prenatal genetic testing has been challenging for the genetic counseling profession to digest because the accusation is at odds with how the profession views itself: Genetic counselors view themselves as "*helping people* understand and adapt to the medical, psychological and familial implications of genetic contributions to disease" (Resta et al. 2006, my italics). The demographic attracted to this profession, typically white, middle-upper class women,[6] frequently believe themselves to be good, tolerant individuals, who respect all choices and celebrate human diversity. The goal, as many genetic counselors see it, is to promote informed choices, not to send messages about the lesser value of certain people. What could be wrong with providing more information? And, as genetic counselor Robert Resta puts it, gently encouraging genetic counselors to laugh at themselves, "Some of our best friends are people with disabilities, right?" (Resta 2011: 1786).

[6] Early genetic counselors were men, but by the 1970s most counselors were white, middle-upper class women; and, according to a 2010 survey by National Society of Genetic Counselors (NSGC) women still make up 95% of the field, with 92% identifying as white or Caucasian (Stern 2012: 25–26). See also p. 74 for a discussion of Nancy Steinberg Warren's 2004 organized retreats on challenges to diversifying the profession.

Does prenatal genetic testing or selective abortion following testing send a "message" that disabled lives are not inherently valuable? A number of authors have raised objections to the *expressivist argument*. Buchanan (1996) argues that there can be no message if the person supposedly sending the message does not actually hold the belief that is the content of the message received. Lindemann Nelson (2000) disagrees, to a certain extent, and offers the example of the confederate flag: The flag may send a message in support of slavery even if the person flying the flag insists that he merely intends to honor his heritage and ancestry. But, says Lindemann Nelson, "abortions are not flags" (2000: 197). An abortion, unlike a flag, is not a symbol or a socially agreed upon way of communicating in the broader context of a language.

Regardless of the ontological status of abortion, one could argue that a message has been received. Bellamo (2009) found that 80% of study participants (377 members of advocacy organizations for Down syndrome) agreed with the statement "Prenatal diagnosis is used to decrease the population of individuals with disabilities," and 90% agreed with the statement "Genetic counselors influence patient decisions." Furthermore, even if counselors do not believe or intend the purpose of prenatal testing to be decreasing the population of individuals with Down syndrome, the fact remains that this is a consequence of prenatal testing. Even though the chance of conceiving a baby with Down syndrome increases with the age of the mother, the number of babies born with Down syndrome has not increased in countries or populations where delayed childbearing has increased (Resta 2011; Cocchi et al. 2010; Caplan 2015; Stern 2012). However, there has been a "relative increase and re-distribution of babies with Down syndrome and other disabilities among certain ethnic and lower socio-economic status groups" (Resta 2011: 1787). Since the chance to conceive a baby with Down syndrome is equal in all populations, the breakdown of births of babies with Down syndrome according to socioeconomic and cultural lines suggests that attitudes towards and/or access to prenatal testing are responsible for this difference (Resta 2011; Cocchi et al. 2010).

Even if genetic counselors do not intend to send a message about the value of life with a disability, empirical studies about information given during genetic counseling sessions support claims by disability advocates that genetic counselors harbor a negative bias toward people with disabilities. Farrelly et al. (2012) found that genetic counselors were more likely to mention termination than pregnancy continuation and adoption. Studies also show that genetic counselors focus primarily on the biomedical aspects of genetic conditions rather than on quality of life issues for a child with a disability (Roter et al. 2006; Farrelly et al. 2012). Counselors did not, for the most part, offer information about what capabilities and skills might be present alongside limitations. In other words, genetic counselors were neither perceived to be offering, nor claiming by self-report to offer, a balance of both positive and negative dimensions of various conditions (Madeo et al. 2011). It should be acknowledged, however, that in response to these findings many genetic counseling programs and genetic counselors have made an explicit attempts to address biases and improve relations with the disability community. Interestingly, responses to the disability critique are generally couched in terms of enhancing client autonomy. It is to these responses that I now turn.

Responding to the Disability Critique

The ideal of non-directiveness[7] has long dominated the genetic counseling ethos, floating intact above the ebb and flow of scholarly critiques about its possibility and desirability (Kessler 1997; Biesecker 1998; Caplan 1993; Gervais 1993; Sorenson 1993; Patterson and Satz 2002; Weil 2003; Arribas-Ayllon and Sarangi 2014; Suter 1998). For the purposes of this paper, the term can be understood in its most simplistic sense: The role of a genetic counselor is to offer her client relevant information about genetic disease and risk, and allow the woman to make her own decision without influencing, i.e. *directing*, her in one way or another. The assumption is that the client can make an autonomous decision only when the counselor behaves in a non-directive manner, and the counselor achieves her goal of non-directiveness if she offers neutral, scientific facts and omits opinions about what the patient ought to do. A neat distinction between facts and values sits unexamined in the background. Genetic information is assumed to be value-neutral, and so if the genetic counselor presents information and does not tell the client how to act on it, she is assumed to be leaving her values out of the discussion, including any value-judgments about the value of life with disability.

The professional motivation for adopting non-directiveness as a foundational principle for genetic counseling is usually described as a means to distance or differentiate genetic counseling from eugenics (Weil 2003; Biesecker 1998; Sorenson 1993). Resta (1997) questions this narrative by pointing out that early practitioners in "human genetics" seemed to have no qualms affirming non-directiveness in one breath and eugenic goals in the next.[8] Still, it does not follow from this that the *intent* was not to give an appearance of a distinction between eugenics and human genetics clinics.

Alexandra Stern (2012) points out that Sheldon Reed, the man who coined the term "genetic counseling" as a "kind of genetic social work without the eugenic connotations" (Reed 1974), was well aware of the importance of distancing the new field of human genetics from eugenics. At the same time, he acknowledged that the two terms were synonyms: "As late as 1979, Reed explained in a lecture that 'our present day use of the term 'human genetics' instead of 'eugenics' may be financially and politically expedient but there is no great philosophical distinction between them'" (Stern 2012: 20). Stern suggests that the field of genetic counseling could be considered "neo-eugenic":

---

[7] For a history of how non-directiveness became central to the profession of genetic counseling see Stern, *Telling Genes,* 2012, especially Chap. 6, where she outlines the convergence of several distinct aspects: The development of client-centered counseling by Carl Rogers; Sheldon Reed's choice of the term client rather than patient as the subject of genetic counseling; the birth of bioethics as a discipline, along with its particular focus on client autonomy as response to past abuses (human experimentation, paternalism, etc. Interesting, the noun "non-directiveness" did not enter the literature until the 1980s, when it became part of a "form of professional identity construction" (144); See also James Sorenson, "Genetic Counseling: Values That Have Mattered," for a discussion of the different values that structured the field as it transitioned from eugenics, to medical genetics, to genetic counseling performed by masters-trained professionals.

[8] See Resta (1997) "Eugenics and Nondirectiveness in Genetic Counseling," p. 256.

> Even if dissimilar to the state-sanctioned eugenics of the past, which entailed forced sterilization and marriage laws, the omnipresent pressure on American women to produce the 'best' or healthiest children possible using available genetic and reproductive technologies resonates with the quest for superior biological fitness and could be considered neo-eugenic (2012: 12).

Even though non-directiveness cannot be, historically speaking, a response to the contemporary disability critique of prenatal getting testing, it can be useful to think of it this way. Disability scholars and activists tend to structure their critiques of prenatal testing in terms that suggest that the field of genetic counseling has not yet attained its ideal of non-directiveness. This critique ranges from pointing out specific lapses to raising more abstract or systemic concerns. First, there are the straightforward and easy to address lapses of the profession's own moral code (for instance, if someone were to actively urge a woman to consider termination). More abstract or systematic concerns can be difficult or impossible to address: For instance, if "the process of prenatal counseling itself presupposes an implicit bias to abort any fetus deemed 'defective'" (Patterson and Satz 2002: 119), it is hard to see what can be done besides abolishing prenatal genetic testing, which is clearly not on the table. Problems related to prenatal testing can be remedied only if there is something specific about the process, other than its existence, that can be altered.

The following is an example of an argument about the existence or nature of genetic testing. Jennings argues that the ground and discourse of the profession is shaped by technology, so the "counseling may be neutral as regards the personal beliefs of the counselor, but it cannot be neutral as regards the very context of genetic technology itself" (2000: 136). Like Patterson and Satz's argument, Jennings' argument highlights the existence of technology as a system of power that structures reality and thus implicitly *directs* the conversation between counselor and client: "Prenatal genetic testing technology shapes choice by in effect making everything into a choice," Jennings writes (2000: 135). Prior to what Abby Lippman has referred to as the "geneticization" of pregnancy, pregnancy could proceed without requiring a decision at every turn; women did not find themselves in the office of a genetic counselor, unclear about why, only to learn that because they are older than 35-years-old, they have the option of learning about conditions, which raise the option, if diagnosed, of raising the option of considering a termination.

Balanced Information and Unconscious Negative Bias

There are also critiques in-between the two extremes above, and it is these to which the genetic counseling community has responded. These related critiques, which I will refer to as the *unconscious negative bias* critique and the *balanced information* critique, proceed roughly as follows:

(a) *Unconscious negative bias critique:* Counselors unconsciously replicate the negative biases of society toward people with disabilities because they have had few interactions with people with disabilities and therefore lack the ability to imagine how good life can be with disability (Parens and Asch 2003).

(b) *Balanced information critique*: Counselors offer a biased picture of life with a disability because they discuss negatives such as medical risk but exclude positives such as unique abilities and quality of life (Stern 2012).

With respect to the second critique, the *balanced information* critique, changes began in the 1960s and 1970s. In the 1970s, groups mobilized to reject negative labels such as mongoloid, retarded, handicapped, dumb, etc. (Stern 2012). The work continues today. In 2006, Brian Skotko, medical geneticist and advocate for people with Down syndrome, wrote a Letter to the Editor to the *American Journal of Obstetrics and Gynecology* to point out that a study published in the journal used words such as "handicap," "risk," and "normal," rather than non-directive options, such as "disability," "chance," and "without disabilities." "In previous studies," he writes, "mothers of children with Down syndrome asked their health care providers to use sensitive language during counseling. We must all be reminded that our words make a difference" (2006: 625–626).

The movement to include positive information has, for better or worse, recently culminated in state legislation, such as Pennsylvania's *Down Syndrome Prenatal and Postnatal Information Act*, also known as *Chloe's Law*. This legislation, named after a little girl with Down syndrome, mandates that positive information be included in counseling sessions about Down syndrome and requires the Department of Health to make available "up-to-date, evidence-based information about Down syndrome that has been reviewed by medical experts and national Down syndrome organizations" (Pub. L. 2450, No. 130, Cl. 35, 2014). In addition to information about the medical facts, parents are to be given resources for support such as phone numbers for national and local Down syndrome organizations.

Setting aside the question of whether laws are the best way to encourage the provision of positive information, the general shift towards balanced information is considered by many to be an improvement. Previously, studies (Skotko 2006; Skotko 2009) have shown that mothers who chose to continue pregnancies after a diagnosis of Down syndrome were not happy with quality of the information given during diagnosis. Skotko found that physicians[9] delivering prenatal diagnoses did not claim by self-report to offer unbiased information: 13% said that they "'emphasize' the negative aspects of DS [Down syndrome] so that parents would favor a termination [and] 10% actively 'urge' parents to terminate" (Skotko 2006: 2362). This is clearly an area where providers could improve.

Dixon (2008) argues that medical professionals do not give sufficient information to women following a prenatal diagnosis of Down syndrome for their decisions to be considered truly informed. He refers to the "failure of non-directive pre-abortion counseling" and suggests a link between this failure, as well as biased information during screening and testing, and the "very high abortion rate for fetuses

[9] It should be noted that these studies include physicians, not genetic counselors, and non-directiveness is not generally considered a cornerstone of physician practice. Some studies show that women were less likely to terminate if they received information about aneuploidy from a geneticist or genetic counselor rather than an obstetrician. See Munger et al. 2007.

diagnosed with Down syndrome" (2008: 3). Thus the concept of non-directiveness is flipped on its head. Whereas in the past "non-directiveness" meant listing medical facts and letting the client decide, today, *only* including medical facts is considered directive (because medical facts are not neutral but negatively biased). Some argue that the positive information movement threatens the ideal of non-directiveness (Caplan 2015), but regardless, it is clear that the movement has had real impact on the way genetic counseling is approached.

In addition to improving the quality and balance of information, the genetic counseling community has worked to increase student and counselor interaction with families raising children with disabilities in order to decrease the unconscious negative bias of genetic counselors. Writing in 2007 about her own experiences counseling clients about Down syndrome, Brasington, a genetic counselor, describes her transition from thinking in terms of the medical model to thinking in terms of the social model of disability. Her transformation was in large part due to interacting with families parenting children with Down syndrome. The expectation that this type of experience might transform genetic counselor attitudes, and accordingly counseling methods, is what informed the curriculum design of the genetic counseling program established by Judith Tsipis in 1992 at Brandeis University. Other programs have followed suit, with some programs including disability studies in their curricula and many offering structured opportunities for students to interact with individuals with disabilities in a non-clinical setting (Madeo et al. 2011; Saxton 2000; Sanborn and Patterson 2014).

Further responses to the *unconscious bias critique* have included professional workshops and meetings that encourage communication between the disability community and the genetic counseling profession. What has come out of these conversations is an intention to increase education about disabilities and "exposure of healthcare providers to individuals with disabilities … to foster awareness" (Madeo et al. 2011: 1782). To facilitate interactions that improve communication between the two communities, the American Board of Genetic Counselors offers continuing education credit for counselors who participate in specific volunteer activities (Madeo et al. 2011).

Decreasing Unconscious Bias in Order to Enhance Reproductive Freedom

The changes discussed above are said to be done in the service of altering the bias of genetic counselors so that they will provide genuinely non-directive counseling that allows clients to make informed choices. This justification is regularly given even when there are numerous other important reasons for the changes. For instance, Asch (2003) devotes the introduction of her "Disability Equality and Prenatal Testing: Contradictory or Compatible?" to a discussion of how prenatal testing followed by selective abortion affects "social institutions beyond the family," "express[es] views that worsen the situation for people who live with disabilities now and in the future," and is at odds with "reforming such institutions as schools [and] workplaces" (2003: 316–317). But she then concludes her introduction by stating her intention in the article as follows: "My concern is to facilitate true reproductive choice for women by urging changes in the way prenatal testing occurs and the rhetoric that surrounds it" (2003: 17).

Changing attitudes and language very well may enhance reproductive choice. But Asch's vacillation between the social and the private may reveal the poverty of a framework that justifies only in terms of freedom of choice. To claim that the goal is primarily to facilitate the reproductive choices of women and couples seems a myopic explanation given some of more systemic concerns raised in the literature, including by Asch herself. Attention to some of the specific issues raised by disability activists and scholars suggests that more reproductive freedom is not precisely what is sought. Notice how Asch's writing displays an ambivalence between the social and the private:

> Despite the symbolic and tangible changes attributable to laws like the Americans with Disabilities Act, the nation's disabled population is still less educated, less employed, less involved in civic life, less represented in the political process, and less influential on the design of products than their numbers warrant (Asch 2003: 331).

And then, ten pages later, the reversion to information and reproductive choice:

> Yet I persist in believing that as part of the *goal of creating such a welcoming society*, we must persuade professionals to *change what they tell prospective parents* about life with disability; convince those parents to learn about how children and adults in today's world survive and thrive; and then *endorse the choices* people make about their reproductive and family lives (Asch 2003: 341, my italics).

She is willing to "endorse the choices of parents," but only after they have been convinced of the value of life with disability. When she writes of the "the goal of creating such a welcoming society," I believe she reveals that while reproductive freedom may be valuable, both in itself and as a means, it is not the end. Framing the issue primarily in terms of enhancing autonomy may be too narrow to capture what is needed.

Zooming Out: Solidarity as Shaping Sensibility

A core feature of a prenatal genetic counseling session is to emphasize the personal nature of decisions related to prenatal genetic testing. When clients ask a counselor which tests she would have or what she would do if she had an abnormal result, the counselor's response tends to center around reminding the client how personal the decision is and how much these decisions vary from woman to woman. These decisions certainly are personal, but, as anthropologist Rayna Rapp notes, "private choices always have pubic consequences" (1988). Similarly, Patterson and Satz raise the question of how genetic counselors should respond to the claim that "the very enterprise of genetic counseling is political" (2003: 120). Further, Jennings reminds us of the "enormous political apparatus of scientific research and testing facilities, to say nothing of the enormous public (whether governmental or corporate) investment and expense that genetic testing technology represents" (2000: 131); and Saxton points out that "[a] woman's individual decision, when resulting from social pressure, or colluding with 'a trend,' has repercussions for all others in the society" (2002: 157).

While I would not go as far as Jennings in concluding that "[i]t is breathtakingly implausible … to characterize the use of genetic testing in obstetric practice in our society as a 'private' act in any sense" (2000: 131), I do think that the focus on enhancing reproductive choice frames the discussion in terms of the private realm. Reproductive decisions certainly are a private matter, but genetic testing, broadly speaking, has social dimensions as well. In this section, I explore how the concept of solidarity might contribute to what Jennings and Dawson refer to as the calculus of consent, or, in this context, the reasons for responding to the call of the disability community. I conclude by considering the contribution of the concept of solidarity for imagining future responses to the disability critique of prenatal testing. I begin with some clarification of the terms calculus of consent, moral imagination, and solidarity.

Jennings and Dawson use the term "calculus of consent" to encompass justifications for obeying laws, rules, and norms that benefit others as well as "stories that a society tells about itself and that individuals tell about their place in it" (Jennings and Dawson 2015: 31). As adapted to my topic, the calculus of consent will refer to justifications of the guidelines and suggestions discussed above, such as "to enhance reproductive freedom" or "to achieve genuine non-directiveness." By moral imagination, Jennings and Dawson mean "the capacity to take a critical distance from the given, to think reality otherwise" (Jennings and Dawson 2015: 31). In the same way, I use moral imagination in the context of this paper to refer to taking a step back from the controversy and contemplating new possibilities for the future. Taken together, I use the concepts of moral imagination and calculus of consent to reflect on the past and imagine the future.

I use the concept of solidarity to refer to a pre-reflective bond between individuals and between grouped individuals that moves us to act on one another's behalf "just because." Or, put more philosophically, it is the condition for the possibility of recognizing one another as ends, but not as abstract ends: As embodied, vulnerable, mutually-dependent beings who stand in a dialectical relation of "needful freedom" with the world (Jonas 2001; Kittay 2011). In other words, our individuality and freedom are premised upon dependence. We cannot go out into the world and pursue our diverse projects as individuals without first attending to the material needs of our bodies. The response to an *other*, in the recognition of this shared dependence, is rooted in a feeling of solidarity.

The Calculus of Consent and the Response to the Disability Critique: Enhancing Autonomy?

Jennings and Dawson suggest that one of the crucial contributions of the concept of solidarity to bioethical discourse is to add a relational dimension to other ethical principles. Without the concept of solidarity, they say, the interests of individuals are viewed as separate and distinct, constituted prior to relationship, not in and through relationship with others. The picture is one of isolated atomic individuals whose interests are sometimes at odds with one another and sometimes compatible, but only incidentally or instrumentally so: "a vision of the individual agent unencumbered, as it were, by solidarity is a vision that stresses the uniqueness of each person and emphasizes difference and separation rather than sameness and commonality. Instrumental ties are the limit of relationally" (2015: 33).

Viewed within an individualistic framework, what is the justification for the genetic counseling profession to work with the disability community? Perhaps it does so out of a kind of "enlightened self-interest": The profession may improve its public image when it demonstrates attention to the accusation of negative bias towards individuals with disabilities. It also improves its "self-image" by proving it can live up to its own ideal of non-directiveness and serve its clients in the way it claims to: Providing information so that clients can make autonomous decisions. This explanation does not seem to capture the reasons why the genetic counseling profession has worked with the disability community or why it should continue to do so.

In a society without solidarity, Jennings and Dawson write, "individuals must obey common rules … not for the sake of others or their rights or interests—those things are incidental—but for the sake of their own protection … There is no encumbrance here, no normative push of commitment or obligation, no motivational pull of mutual recognition and resemblance" (2015: 33). To apply this thinking to the genetic counseling profession's response to the disability community is not to say that genetic counselors are bad people or that they do not want to show respect for individuals with disabilities; I think they do, and this is precisely why framing the "why" of responding to the critiques of above in terms reproductive freedom does a disservice to both communities.

When one considers what the disability community gains from working with the genetic counseling profession, the limits of an individualistic framework, i.e. the reproductive freedom framework, become clearer. As discussed above, some in the disability community say that their aim is to improve reproductive freedom. Arthur Caplan wonders if this is whole motivation of some of the recent positive information laws: "They see the legislation as pro-information and, thus, pro-client autonomy, although, if abortion rates did not change in states with such laws, it is fairly certain the legislation would be seen by many pro-life and disability proponents as a failure" (2015: 3). If this is true, then providing balanced information to enhance the reproductive freedom of women is a means to an end: To decrease abortion rates of individuals with disabilities, or at least to create a welcoming society where individuals with disabilities are respected and able to participate in society to the fullest extent possible. The goal, then, is not merely or even primarily "choice," but, to use the language of Jennings and Dawson, "right recognition" and "right relationship." And the genetic counseling community responds or should respond to this call for recognition out of an obligation, because of the "motivational pull of mutual recognition and resemblance," not simply to enhance the autonomy of clients or improve its perception among the public and in the disability community.

Again, this is not dismiss the importance of reproductive autonomy, but to zoom out and consider the conditions that make individual freedom possible. Jennings and Dawson describe how solidarity grounds the structural context of freedom:

> The normativity of social political life is grounded on what might be called 'right recognition' and 'right relationship.' We define solidarity as a moral practice that is fundamental to a social and cultural structure of right relationship. Right recognition is a condition of moral and political membership, rights, and equality—the recognition of the moral standing and respect of each person. Right relationship is a condition of mutuality—the mutuality of interdependence, care, and concern for others and their relational human flourishing (2015: 32).

Applying these ideas, we would justify the actions of the genetic counseling profession not by saying it is good for the profession or even that it is good for the reproductive freedom of clients; instead, we would look to the background and context of a right like reproductive freedom and say it is good to recognize the moral standing of people with disabilities because, well, they have moral standing and deserve recognition, just as we all do. While people can and have offered instrumental reasons for giving rights to people who lack them (black people, women, LGBTQ individuals, and so on), ultimately, the reason to, for instance, grant black people or women the right to vote, is not that it benefits the country but that they are persons who deserve recognition and the rights that follow from this recognition. Likewise, we stop using offensive language to describe black people, women, gay people, trans people, people with three chromosomes, etc. not to enhance freedom, although this may occur incidentally, but to respect the humanity of the other.

The argument applies as well to the justification for increasing the "exposure" of genetic counseling students and practicing genetic counselors to individuals with disabilities. Why should genetic counseling students interact with people with disabilities? Why should genetic counselors advocate for people with disabilities or otherwise interact with them in non-clinical settings? The argument is that this "exposure" decreases unconscious bias against people with disabilities by enlarging the imagination of counselors to include living well with a disability. This in turn makes non-directive counseling possible because the counselor has supposedly freed herself from bias. So in this view, the justification for meeting people with disabilities is to enhance the autonomy of pregnant women via non-directive counseling.

Some or all of this may be true, but I would argue that the broader reason for interacting with people is less instrumental: It is not "so that" or "in order to," but because this recognition is owed and is long overdue. The demand to recognize the other as fully human is rooted in a relationship that already existed but was un- or under-acknowledged. Recognition is not about interests, desires, or results; it can be inconvenient and uncomfortable to give up a position of power that was premised upon others not being recognized, as is clear from the history of other groups that have struggled for political recognition and social respect. But we have an obligation to do it anyway: "just because" or "because it is owed":

> [I]t is moral commitment, not strategic advantage, that lies at the motivational core of solidarity ... At least in an elementary way, the act of standing up for establishes mutuality and reciprocity among individuals and groups involved in the struggle against injustice (Jennings and Dawson 2015: 36).

When the genetic counseling community stands beside the disability community, when it responds to concerns as best it can, when it works for the equal participation of citizens in our shared society, our shared humanity is affirmed. Individual rights, such as reproductive choice, are possible only in the context of this mutual recognition.

Historical Memory and the Moral Imagination

I have up until now concentrated on what has occurred and how it tends to be explained. I would like to conclude by considering how the concept of solidarity might enrich imaginings for the future, particularly with respect to the education of prenatal genetic counselors. For a topic so intimately and perilously related to individual and social identity as genetics, a sense of temporal context is essential. Jennings and Dawson describe the context-giving dimension offered by a focus on the concept of solidarity:

> Solidarity grows out of a sense of historical memory and tradition, and it feeds on the gratitude that others have made to one's way of life ... [S]olidarity is a concept that inherently leads us to view our actions and the rights, well-being, health, and dignity of others as placed spatially and temporally, as bound together in a here and a now (2015: 32).

Insofar as the role of the genetic counselor is to help a client make sense of genetic information, she must be sufficiently steeped in the history of genetics and its abuses to anticipate the social reverberations of her explanations and respond with nuance and sophistication. Stern refers to the "considerable historical amnesia" of the profession with respect to its "racialized past" (2012: 73). Many communities of color associate the field of human genetics with the forced sterilization of the very recent past.

The pedigree, the mainstay of the genetic counseling session, was developed by early twentieth century eugenicists: "From 1910–1939, the ERO [Eugenics Records Office] as America's premier eugenics research organization, train[ed] the hundreds of eugenic field workers who traveled from state to state to produce pedigrees of supposedly 'defective' families" (Stern 2012: 34). The pedigree then "migrated intact" to medical genetics in the 1940s (Stern 2012: 34) and continues to be used today. Students taking pedigrees should have the opportunity to learn that each line sketched across the page recalls a potent history; they should not be caught unawares by clients who feel the echo of past traumas in the casual line, circle, and square of the pedigree. To add a sense of historical memory to the genetic counseling curriculum would mean a greater emphasis on the origins of human genetics, no matter how shameful—or rather, precisely because of the shameful aspects of this history. As Stern writes, "The field of genetic counseling carries burdensome historical baggage that imposes limitations and can unwittingly hinder the field and its practitioners" (2012: 3). Perhaps the profession should spend less energy distancing itself from eugenics and more energy understanding the eugenic impulse and its temptations. To include greater attention to the history of eugenics and the evolution of the field of human genetics is to acknowledge what Hannah Arendt called "the banality of evil," and to admit that the frailty of good intentions will never be a thing of the past. Nathaniel Comfort refers to the relief of suffering and human improvement as the twin impulses of eugenics and all of medical genetics. Eugenics coupled with state control is only one manifestation of eugenics; the impulse, he says, is "timeless" and manifests differently in different ages: "It is the urge toward selection of the best offspring possible, toward the elimination of hereditary disease, and toward human engineering—'the self direction of human evolution,' as one Progressive-era poster put it" (Comfort 2012: xi). To recognize the way that the eugenic impulse continues to structure our motivations today, and cultivate a deeper appreciation of this continuity, may help genetic counselors better understand how the public perceives the field (Stern 2012). To study history is to recall that the present will some day be the past and to remain curious about what people in the future will conclude looking back.

To add a sense of historical memory to the genetic counseling curriculum might also mean a greater attention to the history of ideas, such as autonomy, nondirectiveness, and value-neutrality. "If any one of us ever did, no one in our group can imagine having a view from nowhere," write Parens and Asch following the completion of the Hasting Center's two-year project exploring the disability critique of prenatal testing (2000: ix). What would it mean to have a "view from nowhere" and what theoretical insights led many philosophers to reject this possibility? While I would not recommend that genetic counseling students delve deeply into debates about scientific realism, I cannot help but find it disappointing that fifty-four years after Thomas Kuhn wrote about scientific observations being theory-laden, and twenty-five years after Dan Brock explained the difficulties with assuming a fact/value distinction in the health care provider/client relationship, genetic counseling is taught as though there were not mounds of literature, some written by genetic counselors, complicating this neat division of labor.

For genetic counselors to be equipped to facilitate constructive conversations with clients who are grappling with complex moral questions, they need to have a sense of the history of debates about moral questions relevant to the field. The uninterrogated acceptance of the fact/value distinction, in which genetic counselors bring the facts and the clients brings the values, leads genetic counselors to a naïve kind of ethical relativism in which client beliefs are true because they are believed by clients. While affirming client values is important, oftentimes clients are struggling to uncover their values or to distinguish their own values from the values of their culture, or their parents, or their partners. In situations like these, mere affirmation, while clearly non-directive, is not particularly affirming of the client's genuine needs.

I want to emphasize that genetic counselors and those who work closely with them have generated a significant amount of literature about the history of the field, including the relevance of applying alternative theoretical frameworks to problems faced by the field. For instance, Satz and Patterson apply feminist standpoint epistemology to the tension between the genetic counseling profession and the disability community. They quote feminist philosopher Sandra Harding putting to work Patricia Hill Collins's African American feminist philosophy on the project of "mak[ing] dominant groups 'fit' to engage in collaborative … enterprises with marginal peoples" (Harding 1993: 68). With respect to improving relations with the disability community, some of the work of becoming 'fit' is accomplished by increasing interactions with individuals with disabilities. Nothing compares to the experience of knowing 'this' person. However, these 'particular' interactions could be even more powerful if students had the opportunity to explore various frameworks within which they could situate their insights. While many genetic counseling students and genetic counselors aim to be accepting of everyone and affirming of all difference, the suggestion here is that being fit to interact with certain people requires going beyond mere intention to accept others:

> Such a project requires learning to listen attentively to marginalized people; it requires educating oneself about their histories, achievements, preferred social relations, and hopes for the future … it requires critical examination of the dominant institutional beliefs and practices that systematically disadvantage them; it requires critical self-examination to discover how one unwittingly participates in generating disadvantage to them and more (Harding 1993: 68).

Rigorous self-examination cannot occur simply by turning inward; genetic counseling students must also have some opportunity, however limited due to the inevitable time constraints of training, to soak up some of the lessons from the decades of discourse on the moral issues related to genetic counseling. To further disability awareness and student self-awareness, we need content that links past and present, that baffles and inspires, that concretizes differences while recalling our shared vulnerabilities as embodied beings.

The actions that the genetic counseling profession has taken thus far in addressing balanced language in genetic counseling sessions and increasing interactions between people with disabilities and genetic counselors may be viewed as the first indication that the framework of solidarity informs, albeit unacknowledged, the decision-making of the genetic counseling community. These changes may indeed enhance reproductive autonomy, but to recognize solidarity as a tacit value (Jennings and Dawson 2015) is to enlarge the framework that structures the moral imagination of the genetic counseling profession in a way that points to the relational underpinnings of autonomy.

To see what was previously hidden opens the possibility of envisioning alternative forms of moral learning. For instance, Jennings and Dawson refer to 'standing up beside' as the fundamental posture of solidarity, and they name three additional aspects that trace the evolution of possible growth: *Standing up for, standing up with*, and *standing up as*. While the distance to the kind of identification indicated by the preposition 'as' is a long way off and perhaps not desirable as a goal, the genetic counseling profession has been working on taking a public stand in response to calls from the disability community. The genetic counseling community has already begun the work of 'standing beside' the disability community: Recruiting individuals to genetic counseling programs and genetic counseling training program advisory boards who have disabilities or have cared for someone with a disability (Shakespeare et al. 2009; Madeo et al. 2011), working with advocacy groups, and getting to know individuals with disabilities outside clinical contexts, are all instances of moving closer, of positioning ourselves in a shared space. To stand beside each other in the future, to become fit for collaboration, is to also to recall the temporal relations that shape present positions. Perhaps we can further the lateral movement already begun by also stepping back, zooming out, and recalling the past as we reflect on what implicit habits of thought structure decision-making and constrain visions for the future.

References

Americans With Disabilities Act of 1990 (1990) Pub. L. No. 101-336, § 1, 104 Stat. 328

Arribas-Ayllon M, Sarangi S (2014) Counselling uncertainty: genetics professionals' accounts of (non)directiveness and trust/distrust. Health Risk Soc 16(2):171–184

Asch A (2000) Why I haven't changed my mind about prenatal diagnosis. In: Asch A, Parens E (eds) Prenatal testing and disability rights. Georgetown University Press, Washington, DC, pp 234–358

Asch A (2003) Disability equality and prenatal testing: contradictory or compatible? Fla State Univ Law Rev 30(2):315–342

Bauer PE (2005) The abortion debate no one wants to have. The Washington Post. http://tony-silva.com/download/abortiondowns-washpost.pdf

Bellamo MA (2009) The perspective of prenatal genetic counseling in the Down syndrome advocacy community. Unpublished Thesis, University of South Carolina, University of South Carolina School of Medicine, Columbia

Biesecker B (1998) Future directions in genetic counseling: practical and ethical considerations. Kennedy Inst Ethics J 8(2):145–160

Brasington CK (2007) What I wish I knew then ... reflections from personal experiences in counseling about Down syndrome. J Genet Counsel 16(6):731–734

Buchanan A (1996) Choosing who will be disabled: genetic intervention and the morality of inclusion. Soc Philos Pol 13(2):18–46

Caplan AL (1993) Neutrality is not morality: the ethics of genetic counseling. In: Bartels DM, LeRoy B, Caplan AL (eds) Prescribing our future: ethical challenges in genetic counseling. Aldine de Gruyter, New York, pp 149–165

Caplan AL (2015) Chloe's law: a powerful legislative movement challenging a core ethical norm of genetic testing. PLoS Biol 13(8):e1002219

Cocchi G, Gualdi S, Bower C, Halliday J, Jonsson B, Myrelid A, Mastroiacovo P et al (2010) International trends of Down syndrome 1993–2004: births in relation to maternal age and terminations of pregnancies. Birth Defects Res, Part A 88(6):474–479

Comfort NC (2012) The science of human perfection: how genes became the heart of American medicine. Yale University Press, New Haven

Dixon DP (2008) Informed consent or institutionalized eugenics? How the medical profession encourages abortion of fetuses with Down syndrome. Issues Law Med 24(1):3–59

Farrelly E, Cho MK, Erby L, Roter D, Stenzel A, Ormond K (2012) Genetic counseling for prenatal testing: where is the discussion about disability? J Genet Counsel 21(6):814–824

The General Assembly of Pennsylvania (2014) Down syndrome prenatal and postnatal education act. Pub. L. 2450, No. 130, Cl. 35. http://www.legis.state.pa.us/cfdocs/legis/li/uconsCheck.cfm?yr=2014&sessInd=0&act=130

Gervais KG (1993) Objectivity, value neutrality, and nondirectiveness in genetic counseling. In: Bartels DM, LeRoy B, Caplan AL (eds) Prescribing our future: ethical challenges in genetic counseling. Aldine de Gruyter, New York, pp 119–130

Gilligan C (1982) In a different voice: psychological theory and women's development. Harvard University Press, Cambridge, MA

Harding S (1993) Rethinking standpoint epistemology: what is strong objectivity? In: Alcoff L, Potter E (eds) Feminist epistemologies. Routledge, New York, pp 49–75

Jennings B (2000) Technology and the genetic imaginary: prenatal testing and the construction of disability. In: Parens E, Asch A (eds) Prenatal testing and disability rights. Georgetown University Press, Washington, DC, pp 196–213

Jennings B, Dawson A (2015) Solidarity in the moral imagination of bioethics. Hastings Cent Rep 45(5):31–38

Jonas H (2001) The phenomenon of life: toward a philosophical biology. In: Northwestern University studies in phenomenology and existential philosophy. Northwestern University Press, Evanston, IL

Kaplan D (2000) The definition of disability: perspective of the disability community. J Health Care Law Pol 3(2):352. http://digitalcommons.law.umaryland.edu/jhclp/vol3/iss2/5

Kenny NP, Sherwin SB, Baylis FE (2010) Re-visioning public health ethics: a relational perspective. Can J Publ Health 101(1):9–11

Kessler S (1997) Genetic counseling is directive? Look again. Am J Hum Genet 61(2):466–467

Kittay EF (2011) The ethics of care, dependence, and disability. Ratio Juris 24(1):49–58

Kittay EF, Kittay L (2000) On the expressivity and ethics of selective abortion for disability: conversations with my son. In: Parens E, Asch A (eds) Prenatal testing and disability rights. Georgetown University Press, Washington, DC, pp 165–195

Lippman A (2003) Eugenics and public health. Am J Publ Health 93(1):11

Lindemann Nelson J (2000) The meaning of the act: reflections on the expressive force of reproductive decision making and policies. In: Parens E, Asch A (eds) Prenatal testing and disability rights. Georgetown University Press, Washington, DC, pp 196–213

Mackenzie C, Stoljar N (2000) Relational autonomy: feminist perspectives on autonomy, agency, and the social self. Oxford University Press

Madeo AC, Biesecker BB, Brasington C, Erby LH, Peters KF (2011) The relationship between the genetic counseling profession and the disability community: a commentary. Am J Med Genet 155A(8):1777–1785

Munger KM, Gill CJ, Ormond KE, Kirschner KL (2007) The next exclusion debate: assessing technology, ethics, and intellectual disability after the human genome project. Ment Retard Dev Disabil Res Rev 13(2):121–128

Noddings N (1984) Caring: a feminine approach to ethics and moral education. University of California Press, Berkeley

Parens E, Asch A (2000) The disability rights critique of prenatal genetic testing: reflections and recommendations. In: Parens E, Asch A (eds) Prenatal testing and disability rights. Georgetown University Press, Washington, DC, pp 3–43

Parens E, Asch A (2003) Disability rights critique of prenatal genetic testing: reflections and recommendations. Ment Retard Dev Disabil Res Rev 9(1):40–47

Patterson A, Satz M (2002) Genetic counseling and the disabled: feminism examines the stance of those who stand at the gate. Hypatia 17(3):118–142

Prainsack B, Buyx A (2012) Solidarity in contemporary bioethics—towards a new approach. Bioethics 26(7):343–350

Press N (2000) Assessing the expressive character of prenatal testing: the choices made, or the choices made available. In: Parens E, Asch A (eds) Prenatal testing and disability rights. Georgetown University Press, Washington, DC, pp 214–233

Rapp R (1988) The power of positive diagnosis: medical and maternal discourses on amniocentesis. In: Michaelson KL (ed) Childbirth in America: anthropological perspectives. Bergin & Garvey Publishers, South Hadley, MA, pp 103–116

Reed SC (1974) A short history of genetic counseling. Soc Biol 21(4):332–339

Resta R (1997) Eugenics and nondirectiveness in genetic counseling. J Genet Counsel 6(2):255–258

Resta R (2011) Are genetic counselors just misunderstood? Thoughts on 'the relationship between the genetic counseling profession and the disability community: a commentary.' Am J Med Genet 155(8):1786–1787

Resta R, Biesecker BB, Bennett RL, Blum S, Hahn SE, Strecker MN, Williams JL (2006) A new definition of genetic counseling: National Society of Genetic Counselors' task force report. J Genet Counsel 15(2):77–83

Roter D, Ellington L, Hamby Erby L, Larson S, Dudley W (2006) The genetic counseling video project (GCVP): models of practice. Am J Med Genet C 142C(4):209–220

Sanborn E, Patterson AR (2014) Disability training in the genetic counseling curricula: bridging the gap between genetic counselors and the disability community. Am J Med Genet A 164(8):1909–1915

Saxton M (1998) Disability rights and selective abortion. In: Solinger R (ed) Abortion wars: a half century of struggle: 1950–2000. University of California Press, Berkeley

Saxton M (2000) Why members of the disability community oppose prenatal diagnosis and selective abortion. In: Parens E, Asch A (eds) Prenatal testing and disability rights. Georgetown University Press, Washington, DC, pp 147–164

Shakespeare T, Iezzoni L, Groce NE (2009) Disability and the training of health professionals. The Lancet 374(9704):1815–1816

Skotko B (2006) Words matter: the importance of nondirective language in first-trimester assessments for Down syndrome. Am J Obstet Gynecol 195(2):625–627

Skotko B, Kishnani PS, Capone GT, and for the Down Syndrome Diagnosis Study Group (2009) Prenatal diagnosis of Down syndrome: how best to deliver the news. Am J Med Genet A 149A(11):2361–2367

Sorenson J (1993) Genetic counseling: values that have mattered. In: Bartels DM, LeRoy B, Caplan AL (eds) Prescribing our future: ethical challenges in genetic counseling. Aldine de Gruyter, New York

Stern A (2012) Telling genes: the story of genetic counseling in America. Johns Hopkins University Press, Baltimore

Stramondo JA (2016) Why bioethics needs a disability moral psychology. Hastings Cent Rep 46(3):22–30

Suter SM (1998) Value neutrality and nondirectiveness: comments on the future directions in genetic counseling. Kennedy Inst Ethics J 8(2):161–163

Wasserman D, Asch A (2006) The uncertain rationale for prenatal disability screening. Virtual Mentor 8(1):53–56

Weil J (2003) Psychosocial genetic counseling in the post-nondirective era: a point of view. J Genet Counsel 12(3):199–211

Wendell S (1996) The rejected body: feminist philosophical reflections on disability. Routledge, New York

Enhancing Future Children: How It Might Happen, Whether It Should

Susan B. Levin

Introduction

According to Savulescu and Kahane's (2009) Principle of Procreative Beneficence (PB), parents are obliged "to create children with the best chance of the best life," which pursuit, as advocates conceive it, includes profound cognitive enhancement.[1] I argue, in contrast, that applying PB thus, should relevant means become available, would deeply harm future children by sealing reason off from the input to moral judgments and decision-making that other faculties provide. In the cases of desire and emotion, enhancement supporters dismiss the nonrational (i.e., what is *other than* reason) as *ir*rational; the relation of each to reason is fundamentally adversarial, making an alliance with reason toward shared ends unthinkable.[2]

Far from seeking harmony between rational and nonrational aspects of us as key to individual and communal flourishing, enhancement supporters laud the boundless elevation of rational ability per se, hinting at an ideal of self-sufficiency. This approach is deeply problematic if, as MacIntyre (1988: 123) observes, "one's rationality [is] not merely supported … but partly constituted by one's membership in and integration into" social milieux. Further, philosophers' customary isolation of

[1] For advocates of the dramatic augmentation of our capacities, particularly regarding cognition and life expectancy, the resulting beings would surpass us so far that they could only be deemed posthuman. For further discussion see Levin (2014, forthcoming).

[2] In this essay, I use "cognitive"/"cognition" and "rational"/"reason" interchangeably because, as far as advocates of cognitive enhancement discuss the matter, reason is in a quite separate compartment from emotion and desire. Further, advocates of cognitive enhancement do not address sense-perception as an aid to the cultivation and use of rationality; this may be related to the fact that they tend to think in terms of ramped-up ability simpliciter, without worrying about either worthwhile ends toward which the faculty might be directed or how the ability is developed, e.g., via experience and education.

S.B. Levin (✉)
Department of Philosophy, Smith College, Northampton, MA, USA
e-mail: slevin@smith.edu

L. Campo-Engelstein, P. Burcher (eds.), *Reproductive Ethics*,
DOI 10.1007/978-3-319-52630-0_3

the cognitive from the noncognitive dimension of human beings (Kipke 2013: 146) is at odds with current research in neuroscience and psychology showing that one cannot "separate emotion and cognition clearly so that we can selectively improve cognitive properties" (154; see also Zohny 2015: 261).[3]

Aristotle's ongoing pertinence to debates about human thought and character is clear not only from his role as the leading inspiration for virtue ethics but also from his bioethical invocation, including, of late, regarding cognitive enhancement. Aristotle's linkage to the debate over cognitive enhancement is a fruitful development, but showing how deeply relevant his ideas are necessitates fuller exploration thereof than has occurred so far. Thus, having critically presented advocates' views in the sections "What Is Cognitive Enhancement?," "The Nonrational Is Misconstrued and Seriously Shortchanged," "Sympathy, Empathy, and Sociality," and "Advocates' Rational Essentialism," I reinforce and deepen that evaluation in the section "Aristotle's Moral Psychology in the *Nicomachean Ethics*" through sustained engagement with Aristotle's account of psychic harmony and moral virtue in that treatise.[4]

Savulescu and Kahane (2009: 289) assert that, to know how to direct enhancement endeavors, "we need to form reasonable opinions on difficult questions about the nature of well-being and the good life." Enhancement supporters have, however, been largely silent on this question, preferring to engage with opponents via a methodology of "overlapping consensus" (Schaefer et al. 2014) that steers clear of such inquiry *by design*. Because the enhancement debate ultimately concerns what values our views of flourishing embody, it should be recast so that this crux is squarely at the fore. As we embark on this reframing, we would do well to bear in mind Aristotle's insights about the nonrational in relation to reason and his unwavering focus on the human "that for the sake of which" (*hou heneka*) all that we do is, perforce, undertaken.[5]

[3] Some of this evidence comes from the study of psychopaths (Kennett and Fine 2008).

[4] Though its focus is quite different, the present paper complements the work of Radoilska (2010), who draws on the *Nicomachean Ethics* when arguing that cognitive enhancement would likely jeopardize the very possibility of agents' excelling in terms of their contributions to epistemic results and appreciation of the epistemic endeavor. In contrast, Fröding's (2011) assertion that cognitive enhancement could "mimic" some "worthwhile aspects of the virtuous life" per Aristotle (229, cf. 232) reflects a failure to appreciate sufficiently the non-negotiability of Aristotelian moral virtues as uniquely exacting excellences, ones requiring balance between rational and nonrational involvement. In addition, Fröding leaves out of account Aristotle's insistence that the subject matter of ethics differs substantially in its degree of a priori exactitude from mathematics (*Nicomachean Ethics* VI 8, cf. I 3), which means, in the former case, that extensive habituation and experience from early childhood are needed for one's development of the requisite contextual attunement (I 4, 7, II 1, III 5).

[5] Though arguing the point falls outside my purview here, I wish to register my view that while Aristotle's own handling of reason and the nonrational in the *Nicomachean Ethics* is rooted in the virtue-based approach that chiefly inspired contemporary virtue ethics, the moral-psychological points it makes about these are logically detachable and quite important whether or not one is committed thereto.

What Is Cognitive Enhancement?

Though advocates assert or strongly imply far-reaching positive effects of cognitive enhancement on well-being, they do not offer a clear and consistent picture of what "cognitive" enhancement would incorporate (cf. Hildt 2013: 4). The following passages illustrate the problem:

> Th[e] relation between autonomy and cognition is generally positive—greater reasoning, deliberation and evaluation typically leads to greater autonomy. There are a number of ways one could cash out the relationship between reasoning and autonomy. … [W]e will focus on three: deductive/logical competence, comprehension (including the avoidance of false beliefs), and critical analysis. (Schaefer et al. 2014: 126)

The focus here is on discursive, or stepwise, reasoning, in particular, logical entailments and exclusions.

Elsewhere, Savulescu (2005: 38) contends that "what separates us from other animals is our rationality, our capacity to make normative judgements and act on the basis of reasons." Our first passage featured a logical process that is "completely content-neutral about values" (Schaefer 2011: 10). But here in the second, "rationality" is equated with an ability to render "normative judgements," with the term "normative" reflecting a necessary tie to content and values in the moral domain that is lacking in Schaefer et al. (2014: 126). Savulescu (2005: 38) continues:

> When we make decisions to improve our lives by biological and other manipulations, we express our rationality and express what is fundamentally important about our nature. And if those manipulations improve our capacity to make rational and normative judgements, they further improve what is fundamentally human. Far from being against the human spirit, such improvements express the human spirit.

"Rational" and "normative" are once again linked, though now—versus the earlier "make normative judgements and act on the basis of reasons"—both terms are used of judgments.

In our final illustration, "intelligence" is said to encompass "many kinds: memory, temperament, patience, empathy, a sense of humour, optimism and just having a sunny temperament" (Savulescu 2005: 37). In this case, traits from memory to empathy are lodged, without elaboration, under intelligence, versus—at least in part and more plausibly—empathy's being located under the nonrational qua emotion and requiring cultivation as such for the sake of its bearing on moral judgment. Intelligence here spans far more territory than that associated with rationality/cognition in the passages treated above. But if the faculty to be augmented under "intelligence" is supposed to differ from what falls under "rationality"/"cognition," no such distinction is made. What is more, "intelligence" and "cognitive capacities" are used interchangeably elsewhere (Schaefer et al. 2014: 130–131). Taken together, the foregoing passages exemplify the fact that enhancement advocates offer an array of statements, not obviously congruent, about what cognition spans and thus what its heightening would encompass.

The Nonrational Is Misconstrued and Seriously Shortchanged

Having briefly documented the relative lack of clarity and consistency in enhancement proponents' constructions of rational ability, I turn to their severance of that capacity from the input to judgments and decision-making that nonrational aspects of us provide. The separation is problematic if, far from being simply an unfortunate feature of human existence as things stand, moral decision-making at its best requires significant involvement of the nonrational. This view, deeply true for Aristotle back in the fourth century BCE, is increasingly confirmed by findings in neuroscience (Borg 2008: 159); indeed, according to McGeer (2008: 229), "our moral nature is shaped by (at least) three different tributaries of affectively laden concern."

How, in defenses of cognitive enhancement, is the nonrational shortchanged?[6] Schaefer et al. (2014: 126) set the temptation to indulge a desire against reason's directive to refrain from its gratification:

> Strictly speaking, desires are non-cognitive. However, reasoning capacities can be crucial in resolving potential conflicts. In the first place, some logical abilities will be needed to recognize that there is a conflict in need of resolution. An unwilling addict ... must recognize the conflict between the lower-order desire to consume some substance and the higher-order desire to cease consumption of that substance (and comprehend the greater importance of the higher-order desire) in order to begin to overcome her condition.

This passage contains unclarities and unargued assumptions. Controversially, the relation of reason to desire is viewed as not just tense but fundamentally hostile such that drawing toward an object of desire, as opposed to holding back, constitutes akratic failure. Further, it is not evident what relation the authors intend between higher-order desires and reason. If reason itself is the source of higher-order desires, what renders them noncognitive? Alternatively, are only lower-order desires thoroughly noncognitive, while higher-order ones possess a cognitive dimension? If they lack a cognitive facet, what, precisely, makes higher-order desires superior noncognitive phenomena, and on what basis, if any, can they coordinate with reason? Whatever the answers, advocates' construction of rational dominance excludes a picture on which the noncognitive qua desire evidently has an optimal relation—one of balance, say—to reason.

Though Schaefer et al. (2014) do not address emotion, so-called negative emotions are treated elsewhere as fundamentally deleterious and hostile to the rational. Per Savulescu and Kahane (2009: 281), we would be better off if they were eliminated: "reproducers also have strong reasons to seek to prevent even an innate tendency to negative affect, or the severe impairment in social skills associated with Asperger's syndrome." Not simply its expression, but our very capacity for strong "negative" emotion, would be removed, as in, genetically deleted.

Here, as with desire in Schaefer et al. (2014), one finds unargued claims. Savulescu and Kahane (2009) take for granted that phenomena like anger function in the disruptive fashion of appetitive desires. They assert the merit of extirpating

[6] The ensuing discussion of desire draws on Levin (2016: 60–61).

the very capacity for "negative affect" as though it were self-evident both what negative emotion encompassed and that it must be subdued to the vanishing point. Savulescu and Kahane fail to provide an account of what makes emotional responses negative versus positive, a necessary basis for which would be a defense of the view that such a bifurcation exists.

Advocates' claims are not only undefended but controversial. Though Haji (2010) speaks of negative emotion, he repudiates the view that negative and positive here are, as such, adversarial. Quite the contrary: "Fear and empathy are different (negative and positive) emotions that together play a necessary role in the capacity for anticipatory guilt and regret. … There is … a fairly intimate connection between fear and empathy, on the one hand, and moral internalization, on the other, and in virtue of this connection, an indirect relation between fear and empathy and ethical perception" (141–142). This means that "deficits in fear, guilt, and empathy … prevent internalization of *moral* norms of conduct … hamper[ing] development of … capacities of *ethical* perception" (143; italics in original).

On complementary grounds, Tappolet (2010: 336), too, rejects the idea of a negative-positive chasm in the sphere of emotion:

> [T]here is reason to think that different types of emotions have a different impact on attention—positive emotions are thought to widen and not to narrow our attentional focus [and] it is certainly plausible to claim that fear narrows the focus of attention. Although this influence is a-rational, it would be a mistake to infer that it necessarily leads to irrationality. Quite … the contrary, it often makes it possible for the agent to focus on what is important.

Far from addressing such depictions, enhancement advocates do not make it evident how or even whether they allow for positive *emotion*. This is not the same as saying that they do not retain a place for what, for instance, empathy, many *consider* positive emotions. The question is whether enhancement advocates view them as such and, if so, how that salutary assessment squares with what they actually say about the noncognitive (see further the section "Sympathy, Empathy, and Sociality").

As for what is rejected, anger and aggressiveness are not differentiated. Thus, Savulescu and Kahane (2009: 284) contend that "[t]he world and the lives of the people in it might be better if everyone were funnier, more intelligent, more empathetic and less aggressive." This lack of distinction matters, for unlike sheer aggressiveness, justified, strong anger might be not only warranted but interlinked with empathy. One may, for example, react thus to a close friend or family member for failing yet again to live up to the standards set by her own laudable values, where one knows how difficult it can be to do that routinely, sees how detrimental this way of existing is to the person's prospects for living well by her criteria, and is all too aware of how highly capable she is of doing better.

Elsewhere, Savulescu (2005: 37) simply fuses anger and aggressiveness:

> Some children would eat [the marshmallow] as soon as the researcher left, others would use a variety of strategies to help control their behaviour and ignore the temptation of the single marshmallow. … Impulse control has also been linked to socio-economic control and avoiding conflict with the law. The problems of a hot temper can include life in prison. Shyness … can greatly restrict a life. … [J]ust having a sunny temperament [where the point of contrast could be anger and/or aggressiveness] can profoundly affect our lives.

The fact that anger and aggressiveness *may* combine, and are noted to do so with some frequency in prisoners (Shniderman and Solberg 2015: 317, 322), is no warrant for conflating anger and aggressiveness or for concluding that strong anger is, as such, problematic.

But the problem here goes beyond this. On advocates' account, strong expressions of emotions such as anger would never be appropriate, hence Savulescu and Kahane's (2009) investment in extirpating the very capacity for such. This controversial position is not flagged as such, let alone defended. As Tappolet (2010: 343) observes, "if you accept that pity or compassion come[s] with altruistic motivations, you must also accept that fear for [or anger toward] others involves altruistic motivations. … [T]he relation of fear [or anger] to action and motivation is complex. Insofar as emotions are perceptions of values, they can inform us about our practical reasons, such as the fearsome [or anger-provoking]." On this line of thought, and per the earlier example involving anger, stamping out the capacity for strong expressions of emotions such as it and fear would diminish relations with close friends and family members insofar as these emotions, like empathy and sympathy, can be prompted by their plights. Further, from a broader human standpoint, to remove our very capacity for strong anger is to eliminate an important source of motivation to ameliorate social injustice.

It is not evident whether enhancement advocates wish to draw a line between acceptable and unacceptable anger and, if so, how they possibly could. Not only are distinctions involving psychic phenomena like anger broadly contested, but even if an agreed-upon conceptual delineation between morally acceptable and impermissible anger were achieved, enhancement proponents cannot reasonably imagine now with any confidence that a capacity for strong anger could be genetically eliminated in future people without our doing away with the capacity for anger altogether.

Sympathy, Empathy, and Sociality

As we have seen, desire and negative emotion are deprecated and set against reason in a deeply adversarial way. Meanwhile, advocates of cognitive enhancement laud empathy, sympathy, and sociality, and mark them for augmentation. Based on PB, "parents would aim to select children with psychological traits that are likely to increase the future child's autonomy—traits such as foresight or self-control, empathy and sympathy" (Savulescu and Kahane 2009: 282). According to Schaefer et al. (2014: 131), "Even if increased autonomy [through cognitive enhancement] remove[s] a handful of options (such as joining a community that disvalues autonomy) from the menu, many more will be opened up (more career opportunities, better management of resources to obtain what one wants, and even greater ability to discern how to integrate into a wide variety of communities, etc.)." Indeed, cognitive enhancement would itself enable one "to develop greater understanding of herself and others" (Savulescu 2005: 38).

Not only are these encomia striking given advocates' sharply critical lens on the nonrational per the previous section, but, once again, the gaps and unanswered questions are highly significant. Discussions of desire and negative emotion by proponents of cognitive enhancement do not stipulate that positive emotions are omitted from condemnation, nor do advocates elsewhere present a different analysis of positive emotion. This omission is glaring since any critique of "negative" emotion necessarily relies, even if tacitly, on a negative-positive contrast.[7] As for the salutary prong of that contrast, if sympathy and empathy are—or at least centrally involve—the nonrational, supporters of cognitive enhancement have not (1) indicated this; (2) used empathy and sympathy to anchor an approving take on anything nonrational; or (3) presented a favorable view of a subcategory of the nonrational that could be applied to these features. If the nonrational is not being condemned outright, this needs to be made absolutely clear, and positive emotions like empathy explicated.

As previously observed, far from unpacking such traits under the head of emotion, Savulescu (2005: 37) lists empathy under "kinds" of *intelligence*, that is to say, under cognition. The same presumably applies to sympathy, which is often taken to imply less familiarity with particular individuals than empathy. But if cognitive enhancement is itself to augment sympathy and empathy, proponents do not articulate, let alone defend, a view of the mind on which heightened cognitive ability might be expected to intensify those qualities.

Absent further argument, it is highly problematic for advocates to claim that cognitive enhancement would upgrade empathy while at the same time eliminating "an innate tendency to negative affect" (Savulescu and Kahane 2009: 281). On the one side, if supporters take empathy itself to fall under cognition, meaning that cognitive enhancement would include its augmentation, they have not argued for this construction of it. On the other, if empathy—like desire and negative emotion—is itself noncognitive, then it is most difficult to see how cognitive enhancement as such would heighten it. More fundamentally, if it is noncognitive, enhancement supporters have offered us no reason to think that it is *worth* retaining, let alone heightening, because the sole template offered for the noncognitive presents it as warranting quashing and ultimately elimination. Further, enhancement proponents are not off the hook if they claim that empathy itself includes both rational and nonrational dimensions, as they would have to defend this construction of it. In addition, they must show what the cognitive aspect of empathy is such that cognitive enhancement could reasonably be expected to heighten it and why empathy is exempt from advocates' overt, highly critical handling of desire and negative emotion.

As to sociality—which itself requires a live capacity for sympathy and empathy—Schaefer et al. (2014: 126–27) stress the harmful impact of excessive conformity with others' views. While it is clear what enhancement supporters reject in terms of humans' reciprocal impacts, it is not at all evident what brand of sociality they wish to replace it with (see further "*Phusis* (Nature) *1* Carefully Distinguished from *Phusis 2*," the final subsection of "Aristotle's Moral Psychology in the *Nicomachean Ethics*").

[7] On contrast-dependency in human thought and experience, see further Levin (2014: 6, 9–10).

Advocates' Rational Essentialism

As we have seen, when addressing the nonrational as such, proponents of cognitive enhancement do not consider the possibility of real calibration in our expression of nonrational features individually (e.g., anger) or of harmony between rational and nonrational dimensions, where both are deemed essential. Instead, they are concerned to move us to a plane of cognitive operation where the nonrational would not intrude since it had been genetically edited out or because, at minimum, cognition had become so powerful that nonrational impetuses would effectively not register.

Enhancement advocates' routine extolling of cognitive ability and its radical augmentation, with their disparagement of the nonrational when handling it expressly, suggests a leaning toward rational essentialism (see especially Savulescu 2005: 38). Aristotle clearly endorses a version thereof. Contra enhancement supporters, however, he does not simply critique deficient nonrational expressions but foregrounds a key place for the nonrational in human flourishing itself. My challenge in this essay is thus not to rational essentialism of whatever kind but rather to its interpretation by proponents of cognitive enhancement. Aristotle's *Nicomachean Ethics*, to which I now turn, offers a rich account not only of how we humans often go wrong moral-psychologically but also of how nonrational and rational dimensions of us are interwoven in a flourishing life.

Aristotle's Moral Psychology in the *Nicomachean Ethics*

This section addresses the following topics: (a) Aristotle's framework for approaching the nonrational; (b) his union of desire and emotion under the head of *pathê*; (c) the Doctrine of the Mean; (d) Aristotle's notion that emotional responses deemed negative by enhancement advocates may be not simply permissible but morally required; (e) the moral import of pleasure and desire; and (f) Aristotle's delineation of what I call *Phusis* (i.e., nature) *1* from *Phusis 2*.

Grounding Aristotle's Approach to the Nonrational

Aristotle's *Nicomachean Ethics* illustrates how one's account of human nature and flourishing can give pride of place to reason while offering a multi-faceted, rich picture of our psychic operations. In *On the Soul* (II 1–2), Aristotle distinguishes living from nonliving entities on account of the former's possession of soul. Further, he divides living things into basic types due to the kind of soul each has. The psychic capacities (*dunameis*) that Aristotle identifies (II 2–3) are (1) a nutritive faculty that enables growth and reproduction; (2) sense-perception; (3) desire; (4) locomotion;

and (5) rationality.[8] Possession of (1) distinguishes plants from natural, inanimate entities (e.g., fire), while (2)–(4) differentiate nonhuman animals from plants. As *On the Soul* (II 3) makes clear, the capacities are hierarchically arranged such that living things on a higher tier possess not merely their type's distinctive feature(s) but also any feature(s) characteristic of entities lower down. Thus, humans' distinctive capacity is rationality (cf. *Nicomachean Ethics* I 7), but they also possess abilities (1)–(4).

Early in the *Nicomachean Ethics*, Aristotle states that he will address the soul with flourishing (*eudaimonia*), specifically, in view (I 13, 1102a23–24). In keeping with that focus, the treatise attends substantially to the nonrational, specifically, emotion and desire. Though not intrinsically rational, they (unlike the nutritive capacity of all living things) are amenable to rational governance (1102b28–31) such that, in a morally virtuous person, "everything is in harmony with reason."[9]

In theory, reason can be dominant in ways that do or do not require harmony with what is other than it. The latter scenario obtains when rational governance involves the relentless, full subordination of the nonrational; based on the section "The Nonrational Is Misconstrued and Seriously Shortchanged," this seems to be the view of enhancement advocates. Aristotle's picture of admirable conduct and motivation, in contrast, requires harmony between reason and the nonrational (III 12, 1119b15–16). While both Aristotle (III 8, 12) and enhancement advocates recognize that faculties besides reason are inadequate guides to action on their own, only Aristotle sees that other faculties can yet be essential, not merely in the ordinary course of events but for *eudaimonia*. Thus, enhancement advocates' depreciatory lens on the nonrational as such is, at minimum, underdetermined by the evidence that they have provided thus far. This should be all the more concerning to them given recent findings in neuroscience and psychology that favor Aristotle by showing emotion and desire to figure importantly in moral motivation (McGeer 2008: 246–247).

Desire and Emotion United as Pathê

Pathos (plural, *pathê*) is Aristotle's collective term for emotion and desire. *Pathê* include appetitive desire, fear, anger, affection, longing, and pity, "and in general anything that is followed by pleasure or pain" (*Nicomachean Ethics* II 5, 1105b21–23). Together with capacities and characteristics, *pathê* comprise what exists in the soul (1105b20).

[8] Because Aristotle is a monist, "psychic" should not be construed disjunctively with "physical," or "biological." While he separates soul and body for analytic purposes, in terms of beings' actual existence, "we can dismiss as unnecessary the question whether the soul and the body are one: it is as though we were to ask whether the wax and its shape are one" (*On the Soul* 412b6–7; trans. Smith 1984 [line numbers are from the edition of Ross 1956]).

[9] Unless otherwise noted, translations of the *Nicomachean Ethics* are from Ostwald (1962), with certain adjustments; for the Greek, I use Bywater (1894).

Like enhancement advocates, Aristotle puts the nonrational, comprised of desire and emotion, under one collective head. But his account shows that doing so need not commit one to a sharp devaluing of the nonrational that would squeeze out the very possibility of psychic balance, rendering the notion itself unintelligible. For Aristotle, versus devotees of cognitive enhancement, deeming an aspect of us nonrational does not itself signify that the item thus labeled has nothing potentially fruitful to do with reason itself; in the case of *pathê*, "nonrational" picks out what is not intrinsically rational yet able to coordinate and collaborate with reason. In fact, though not rational in their own right, qua psychic phenomena, *pathê* "belong to humans no less than reason does" (III 1, 1111b1).[10]

Aristotle's Doctrine of the Mean and Notion of Psychic Balance

Aristotle's Doctrine of the Mean offers valuable guidance on how to differentiate among aspects and expressions of the human psyche in degree and kind. From start (I 6, 13, II 1–3) to finish (X 9), Aristotle's concern in the *Nicomachean Ethics* is not the sheer subordination, let alone extirpation, of our capabilities besides reason proper but attaining balance among them—crucially including harmony of aims.

Aristotle's view of *pathê* in relation to reason is elaborated through his Doctrine of the Mean, according to which "we can experience fear, confidence, desire, anger, pity, and generally any kind of pleasure and pain either too much or too little, and in either case not properly. But to experience all this at the right time, toward the right objects, toward the right people, for the right reason, and in the right manner—that is the mean and the best course, the one that is a mark of virtue" (II 6, 1106b18–23). Crucially, the mean comprising virtue "involves both *pathê* and actions" (b24–25). While sympathy and empathy per se are not listed, in terms of Aristotle's moral psychology, they belong under *pathê*.

Because the moral mean is categorically different from the arithmetical variety, ascertaining what conduct expresses that mean cannot be routinized but instead often requires fine-grained contextualization (II 6; see also I 6, II 2–3, 5–6, 9). As regards the nonrational, features of situations into which one might enter both impact whether one responds and help to calibrate one's response under the pertinent head (e.g., anger, fear); regarding the latter, the issue for Aristotle is not simply what reactions are morally permissible but which ones, particularly involving emotion, may be morally required (see further the subsection "Strong Emotional Responses Can Be Morally Required").

Cognitive enhancement would decimate the very possibility of subtle calibration, for, per advocates, the preeminent rational preference involving the psychic domain just *is* for greater rational capacity. Seen from within Aristotle's own rational essentialism, in contrast, this preference is not rational because, if acted upon, it would throw off a delicate and important balance among multiple facets of who we

[10] My translation.

are. According to Bostrom (2009: 130), Aristotle's rational essentialism "is plainly not a promising objection to [Bostrom's own transhumanist vision] since it would be perfectly possible for a posthuman to realize a telos of rationality as well as a human being could. In fact, if what is good for us is to develop and exercise our rational nature, this implies that it would be good for us to become posthumans with appropriately enhanced cognitive capacities." Bostrom's confidence is misplaced since, for Aristotle, augmented rational ability is not a goal in its own right. Instead, that capacity is meaningful only when actualized and instantiated in light of a rich, articulated notion of flourishing that is the ultimate telos, or "that for the sake of which" (*hou heneka*), of everything we humans do.

My point here is not that enhancement advocates must adopt Aristotle's own, virtue-based, account of *eudaimonia*. Rather, I wish to reinforce the point that their rational essentialism is only distantly related to his and to emphasize, by drawing on him, that rational capacity itself is bereft of meaning (and potentially quite dangerous) in the absence of a conception of the "why" whose content one expressly formulates and defends. Yet, far from doing this, advocates treat their non-provision of such as an asset of their accounts (see, e.g., Savulescu and Kahane 2011; Schaefer et al. 2014).

Further, rather than speaking of "rationality" simpliciter, as do enhancement advocates, Aristotle identifies and interrelates contemplative, or theoretical, and practical varieties. Why might this matter? As we saw in the section "What Is Cognitive Enhancement?" Savulescu (2005: 38) ties "rationality" to "our capacity to make normative judgements." Reference to Aristotle suggests that what advocates of cognitive enhancement propose would minimize and perhaps eliminate our ability to make those judgments. For him, moral expertise requires exemplary attainment in both types of reason: Not only is there "no practical rationality [i.e., *phronêsis*] … without the virtues of character" (MacIntyre 1988: 136), but both are entwined with virtuous agents' contemplative grasp of human flourishing and specific universals such as courage (Sorabji 1980: 205, 207; *Nicomachean Ethics* 1151a16–17).

Even where there is overlap between portrayals of reason by Aristotle and advocates of cognitive enhancement, there are salient differences. To illustrate: adeptness in stepwise reasoning is important for both (Schaefer et al. 2014; *Nicomachean Ethics* VI 7), but only Aristotle sees clearly that this facility is not valuable in a vacuum but instead proves its mettle in applications that call for marked contextual attunement. Further, reason, whether theoretical or practical, is insufficient to produce activity; for this, desire is always required (*On the Soul* III 10).

Balance, or harmony, exists only in relationships, ones in which all salient threads are suitably aligned. Aristotle's Doctrine of the Mean is a powerful and enduring illustration of such a position. When, in contrast, a view stipulates that threads are to be kept wholly separate, this is normally because its partisans deem one or more of these inferior such that they can overpower or pollute the superior factor. But what if a capacity of which it is stipulated, "Keep it separate because it is fundamentally superior," depends for its richest manifestations on a version of what—here, something nonrational—one believes it must be kept rigidly separated from? Aristotle's handling of anger, to which I now turn, shows what an embrace of enhancement supporters' agenda stands to jeopardize with regard to future children.

Strong Emotional Responses Can Be Morally Required

For Aristotle, unlike enhancement supporters, strong emotional responses are sometimes not just allowed but morally requisite. Given its importance both to advocates of cognitive enhancement (see the section "The Nonrational Is Misconstrued and Seriously Shortchanged") and to him, anger offers a good case study. For Aristotle, unlike Savulescu and Kahane (2009), "our condition in relation to anger is [not] bad" simply because the anger is strong (*Nicomachean Ethics* II 5, 1105b26–27). Instead, we err "if our anger is either too violent or not violent enough. … A man does not receive praise for being frightened or angry, nor blame for being angry pure and simple, but for being angry in a certain way" (1105b27–1106a1). Of course, getting angry as such is no achievement (1106a2–3; II 9, 1109a26–27), but evincing anger "to the right person, to the right extent, at the right time, for the right reason, and in the right way is no longer something easy. … It is for this reason that good conduct is rare, praiseworthy, and noble" (1109a28–30).[11]

What is more, "in some cases we must (*dei*) be angry" (III 1, 1111a30–31). "Not being driven by emotion" (IV 5, 1125b34–35) thus does not mean avoiding strong responses involving emotions simply because, for instance, those not well poised to judge might condemn them as excessive. Quite the opposite, for not just excess but "deficiency … receives blame. For those who do not show anger at things that ought to arouse anger are regarded as fools; so, too, if they do not show anger in the right way, at the right time, or toward the right person" (1126a3–6). On Urmson's (1980: 161) useful elucidation of Aristotle's view, one

> whose character is such that he feels only mild annoyance at a trivial slight and is enraged by torture has a character that is in a mean between one that exhibits rage on trivial as well as important occasions and one that can coolly contemplate the greatest outrages. … To diverge from the mean in the direction of deficiency is as much not to experience and exhibit emotions at all when one should, or not about matters about which one should, or not toward people toward whom one should as it is to exhibit the emotions to the wrong degree.

Considering the matter against the backdrop of Aristotle's view, one risk of our going full steam ahead with cognitive enhancement is that we may edit out of existence in future children the very capacity for powerful, justified anger at injustices, including inequities in access to resources and opportunities that would themselves likely be intensified by cognitive enhancement measures. We also stand to jeopardize the possibility of deep friendship, where the parties "become better people as they are active together and correct one another" (*Nicomachean Ethics* IX 12, 1172a11–12; cf. VIII 8, 1159b6–7), insofar as strong anger can be not only permitted but downright called for by the nature of that tie. An eventual exclusion of the nonrational could well encompass desire and pleasure, too, which Aristotle closely relates as motivators and measures of agents' standing and actions' moral worth.

[11] This is not to say that any and all acts and responses may at times be appropriate, for, as MacIntyre (1988: 121) observes, on Aristotle's account (II 6), certain acts (e.g., adultery) and responses (e.g., *Schadenfreude*) are categorically precluded.

The Moral Richness and Import of Desire and Pleasure

Though far from being a hedonist, Aristotle foregrounds the import of pleasure and pain well beyond a focus on them as direct indicators that something is appetitively desirable or the opposite: "Pleasure and pain are a consequence of every emotion and of every action. … [Thus] virtue has to do with pleasures and pains" (II 3, 1104b14–16). Further, pleasure is not merely an experiential consequent, for Aristotle identifies "the noble, the useful, and the pleasant" as the motivators of decision and choice (b31). Pleasure itself is not condemned; rather, moral assessment of its presence in any situation depends on "whether we feel joy and pain in the right or the wrong way" (1105a7).[12]

Further, "an activity is increased by the pleasure proper to it. … Each activity determines its own proper pleasure. The pleasure proper to a morally good activity is good, and that proper to a bad activity is bad" (X 5, 1175a30–31, b26–28). The same situation obtains with desire (b28–29; VI 2, 1139a29–31). Through observations such as these, Aristotle points to the moral complexity and import of the noncognitive qua desire and thus pleasure.[13]

The constricted vision of enhancement supporters regarding desire (and so-called negative emotion) is closely related to their conflation of two senses of "nature" (*phusis*), which is where their divergence from Aristotle involving the human comes to a head.

Phusis *(Nature)* **1** *Carefully Distinguished from* **Phusis 2**

Aristotle pointedly differentiates between what I call here *Phusis 1* (potentiality)—"nature" qua a capacity not yet developed or defectively so—and *Phusis 2* (actuality): "nature" in the sense of a well-developed ability that is actively deployed.[14] One can oppose his picture to the rational essentialism of enhancement advocates, where a *Phusis 1*-notion of the nonrational appears to exhaust that dimension of us. Their *Phusis 1*-view of the nonrational is coupled with a vague *Phusis 2*-lens on

[12] Aristotle illustrates this scenario via the character of Neoptolemus in Sophocles' tragedy *Philoctetes* (VII 2, 9).

[13] Unlike Aristotle, Schaefer et al. (2014) leave unaddressed the question of whether they view pleasure in lockstep with desire or in some other, perhaps more flattering, way.

[14] When arguing that technological enhancement is needed, whose provision would allegedly help people become virtuous in the vein of Aristotle, Fröding (2011: 226, 231) does not take adequate account of *Phusis 2*'s distinctness from *Phusis 1*. For fine-grained consideration of the distinctions involved in *Phusis 1* and *Phusis 2*, see *Nicomachean Ethics* I 7, II 1–2, VI 8, 11–13, X 6; *On the Soul* II 5.

rationality that comprises a heightening of capacity, in principle limitless, apart from any substantive notion of worthwhile aims (cf. Levin 2016: 61–62).[15]

When they denigrate the nonrational, enhancement advocates fail to distinguish adequately between "nature" by potential versus in actuality. As Aristotle observes, "the end that appears [good] to a particular person … is not simply given to him by nature but is to some extent due to himself" (*Nicomachean Ethics* III 5, 1114b16–17; see also II 1, 5). Thus, the goal one sets for oneself matters greatly. For Aristotle, the pinnacle of rational attainment is the actively virtuous existence in which flourishing consists. Said existence represents the culmination of a protracted development from *Phusis 1* to *Phusis 2*, where the virtuous capacity we have on account of our humanness has been developed into "virtue in the full sense" (VI 13, 1144b14; cf. b16–21).[16] Over and above the sustained cultivation of virtuous characteristics, "virtue in the full sense" requires our enactment of those features: Otherwise, one who possessed virtuous characteristics (*hexeis*) could be said to flourish even if he was "inactive all his life" (I 5, 1095b33). Aristotle rejects this view: "Just as the crown at the Olympic Games is not awarded to the most beautiful and the strongest but to the participants in the contests … so the good and noble things in life are won by those who act rightly" (I 8, 1099a3–7).

By Aristotle's lights, enhancement supporters' claim about the nonrational as disruptive applies only to cases where self-discipline is sorely lacking, that is to say, where attainment is remote even from a well-developed virtuous capacity. Advocates seem to presume that humans just *are* akratic,[17] whereas, for Aristotle, what they reject simply reflects *Phusis 1*, namely, it falls under the head of human potential un- or deficiently realized. The picture of enhancement supporters thus ignores a salient distinction, drawn by Aristotle, between morally strong (enkratic) and morally weak (akratic) persons: It is to the latter alone that "we must attribute … a condition similar to that of men who are asleep, mad, or drunk" (VII 3, 1147a17–18).

Further, surpassing moral strength is self-control proper (i.e., the virtue of *sôphrosunê*), whose possessor does not merely do what is morally required but desires to do so and enjoys acting thus. Because advocates of vigorous enhancement fail to demarcate *Phusis 1* from *Phusis 2*, they presume identified flaws to reflect humanity as such, and this, in turn, renders the idea of a coming-to-fruition from within the human itself a nonstarter.

---

[15] The same criticism applies where transhumanists emphasize artificial intelligence instead of genetics (see Levin forthcoming).

[16] On Aristotle's broader formulation of the point in the *Physics*, *Phusis 2* "is the end or that for the sake of which (*telos kai hou heneka*)" (II 2, 194a28)—the end being "what is best (*beltiston*)" (II 3, 195a24; trans. Hardie and Gaye 1984 [line numbers are those of Ross 1950]).

[17] On Aristotle's account (*Nicomachean Ethics* VII 4), *akrasia* applies to *pathê* generally, not only to appetitive desire. Schaefer and Savulescu (2016) appear to concede indirectly that cognitive enhancement alone may not handle *akrasia* when, having singled out the moral "useful[ness]" of "logical competence" (4), they grant that "akrasia reduction was not included in our present framework … because it does not easily fit with our focus on judgments" (4n5).

What is more, enhancement supporters' praise of augmented sociality notwithstanding, Aristotle's *Phusis 1–Phusis 2* composite requires ties to others of the sort that their accounts cannot obviously accommodate and may preclude. For Aristotle, flourishing—indeed, the very possibility of our reaching *Phusis 2*—necessitates a rich communal setting (*Politics* I 2 [Ross 1957]; *Nicomachean Ethics* VIII–IX, X 9). Interpersonally, friendship (*philia*) is essential to a virtuous life (*Nicomachean Ethics* VIII 1), for "a man of high moral standards will need people to whom he can do good" (IX 9, 1169b13). An active mutuality importantly distinguishes *philia* from mere good will (*eunoia*; IX 5, 12). Such is this mutuality that the parties' contributions are not neatly differentiable, occurring in a context where what happens is often meaningful only if one takes the specifics of that relationship into account.

Enhancement advocates identify what the heightening of cognitive ability would supposedly eliminate from human connections (see, e.g., Schaefer et al. 2014). But they leave it unclear how others would necessarily matter in a fruitful way if dramatic cognitive enhancement, alongside the profound diminishment or elimination of our biological frailty, enabled us to approach the self-sufficiency whose prospective attainment enhancement supporters often esteem. We can see more readily how we (or our successors) would be far less entwined with others than most of us could fathom. To that extent, in antiquity, enhancement supporters' view fits best with that of the Stoics (Diogenes Laertius, *Life of Zeno* [Hicks 1931]; Epictetus, *Encheiridion* [White 1983]), who condemn strong emotional responses—from anger and grief through love—as such, distinguishing them from "good emotions" (*eupathê*; *Life of Zeno*, VII 116–19). The latter are not directed toward individuals but rather to humans in aggregate or, at their most specific, toward others based on their roles in relation to us (e.g., filial respect, owed by children to their parents due to the latter's function as such; *Encheiridion* 30). Because these responses are tepid and generic, what makes *eupathê emotional* responses for Stoics is elusive. Something in this vein is what "augmented" sociality might look like if robust cognitive enhancement were pursued.

Where Do We Go from Here?

The foregoing assessment of arguments for cognitive enhancement, both directly and through the lens of Aristotle's thought, problematizes the notion of heightening "rationality" as far as technically possible apart from (1) the nonrational and (2) an express commitment to substantive views of the values and aims that rationality is intended to serve. These two factors are strongly connected, for one cannot distinguish justified from unwarranted manifestations of psychic capacities like anger if one lacks meaningful notions of the good, just, and so on (on the imperative to investigate these, see Jotterand 2011: 5, 7–8). Absent these notions, how could one articulate and defend violations of the norms represented thereby?

Advocates of cognitive enhancement do not provide the requisite accounts, nor is it evident how, as things stand, they could. For doing so would require toning down if not abandoning their "overlapping consensus" approach (Schaefer et al. 2014; Savulescu and Kahane 2009) when trying to convince critics that there is more common ground than meets the eye between their core values and advocates' own. That approach gets going only if one severs constructions of concepts like rationality, autonomy, and virtue from the theoretical settings—centrally including views of worthwhile aims—within which they have their distinctive meanings. In this process, inevitably, the operative notions will be watered down. The views of enhancement supporters are of course no exception here, being anchored firmly in commitments that inevitably recede from center stage when they seek to highlight what they claim to be shared terrain (on this topic, see also Levin 2016: 60, 62).

As MacIntyre (1988) reminds us, we cannot discuss goods or rationality in a vacuum, for there just *are* bedrock divergences. Further, "disagreements between fundamental standpoints are in key part over how to characterize those disagreements. There is … no neutral mode of stating the problem, let alone the solutions" (144). This means, crucially, that "[p]rogress in rationality is achieved only from a point of view" (ibid.).

Thus, the perspective we adopt on rationality and its ties to the nonrational will markedly impact the future, including the moral standards by which decisions about progeny-to-be will be assessed. Because, under the highly controversial terms of PB, enhancement would itself be morally required, the stakes could not be higher.[18] Once the veil and refuge of "overlapping consensus" are removed, advocates are faced squarely with the tough challenge of defending their controversial and problematic views about cognition and the noncognitive, several of whose shortcomings I have illuminated here.

The matter of where enhancement advocates and their "conservative" critics *do* concur is best approached apart from the distortional frame of overlapping consensus. Here are two such points: "we need to form reasonable opinions on difficult questions about the nature of well-being and the good life" (Savulescu and Kahane 2009: 289), and "[u]nless we begin to understand what is good and ought to be promoted and what is bad and ought to be prevented, we will be in no position to evaluate [the] rapidly advancing scientific possibilities" (Savulescu 2003: 25). These statements imply broad concurrence with Aristotle that nothing can be a meaningful, guiding aim—or "that for the sake of which"—if it is merely a *placeholder* for whatever content one with radically heightened cognitive ability might give it.

For our own and our children's sakes, such statements by enhancement supporters of the need to articulate substantive notions of "well-being and the good life" must not remain unfulfilled promissory notes. The question we should address more concertedly is, for the sake of what, if anything, could the pursuit of vigorous cognitive enhancement be justified? Because the controversy over enhancement is ultimately about values as reflected in aspirations and ideals, reframing the debate to foreground this fact would itself be a marked advance. We need not endorse the particulars of Aristotle's stance. But in this enterprise of recasting, his nuanced handling of

[18] See further Levin (2016).

the nonrational in relation to reason is well worth bearing in mind. In addition, his *Nicomachean Ethics* vividly reminds us that no capacity of ours is meaningful unless it is framed, developed, and enacted in light of a rich notion of individual and communal flourishing that is the ultimate telos, or reference point, of all we humans do.

Acknowledgments I wish to thank audience members at the conference for their feedback on the presentation given there. In addition, I am grateful to Fabrice Jotterand, Jill de Villiers, Julie Ward, and Katie Wing for valuable input.

References

Borg JS (2008) Impaired moral reasoning in psychopaths? Response to Kent Kiehl. In: Sinnott-Armstrong W (ed) Moral psychology, vol 3: The neuroscience of morality: emotion, brain disorders, and development. MIT Press, Cambridge, pp 159–163

Bostrom N (2009) Why I want to be a posthuman when I grow up. In: Gordijn B, Chadwick R (eds) Medical enhancement and posthumanity, international library of ethics, law and technology, vol 2. Springer, Dordrecht, pp 107–136

Bywater I (ed) (1894) Aristotle, Ethica Nicomachea. Clarendon Press, Oxford

Fröding BEE (2011) Cognitive enhancement, virtue ethics and the good life. Neuroethics 4(3):223–234

Haji I (2010) Psychopathy, ethical perception, and moral culpability. Neuroethics 3(2):135–150

Hardie RP, Gaye RK, trans. (1984) Aristotle, Physics. In: Barnes J (ed) Complete works of Aristotle, vol 1. Princeton University Press, Princeton, pp 315–446

Hicks RD, trans. (1931) Life of Zeno. In: Diogenes Laertius, Lives of eminent philosophers, vol 2. Loeb Classical Library. Harvard University Press, Cambridge, pp 110–262

Hildt E (2013) Cognitive enhancement—a critical look at the recent debate. In: Hildt E, Franke AG (eds) Cognitive enhancement: an interdisciplinary perspective. Springer, Dordrecht, pp 1–14

Jotterand F (2011) 'Virtue engineering' and moral agency: will post-humans still need the virtues? AJOB Neurosci 2(4):3–9

Kennett J, Fine C (2008) Internalism and the evidence from psychopaths and 'acquired sociopaths'. In: Sinnott-Armstrong W (ed) Moral psychology, vol 3: The neuroscience of morality: emotion, brain disorders, and development. MIT Press, Cambridge, pp 173–190

Kipke R (2013) What is cognitive enhancement and is it justified to point out this kind of enhancement within the ethical discussion? In: Hildt E, Franke AG (eds) Cognitive enhancement: an interdisciplinary perspective. Springer, Dordrecht, pp 145–157

Levin SB (2014) Transhumanism and enhancement. Wiley, Chichester. http://www.els.net

Levin SB (2016) Upgrading discussions of cognitive enhancement. Neuroethics 9(1):53–67

Levin SB (forthcoming) Antiquity's missive to transhumanism. J Med Philos

MacIntyre A (1988) Whose justice? Which rationality? University of Notre Dame Press, Notre Dame

McGeer V (2008) Varieties of moral agency: lessons from autism (and psychopathy). In: Sinnott-Armstrong W (ed) Moral psychology, vol 3: the neuroscience of morality: emotion, brain disorders, and development. MIT Press, Cambridge, pp 227–257

Ostwald M, trans. (1962) Aristotle, Nicomachean Ethics. Macmillan, New York

Radoilska L (2010) An Aristotelian approach to cognitive enhancement. J Value Inq 44(3):365–375

Ross WD (ed) (1950) Aristotle, Physica. Clarendon Press, Oxford

Ross WD (ed) (1956) Aristotle, De Anima. Clarendon Press, Oxford

Ross WD (ed) (1957) Aristotle, Politica. Clarendon Press, Oxford

Savulescu J (2003) Human-animal transgenesis and chimeras might be an expression of our humanity. Am J Bioeth 3(3):22–25
Savulescu J (2005) New breeds of humans: the moral obligation to enhance. Ethics Law Moral Philos Reprod Biomed 1(1):36–39
Savulescu J, Kahane G (2009) The moral obligation to create children with the best chance of the best life. Bioethics 23(5):274–290
Savulescu J, Kahane G (2011) Disability: a welfarist approach. Clin Ethics 6(1):45–51
Schaefer GO (2011) What is the goal of moral engineering? AJOB Neurosci 2(4):10–11
Schaefer GO, Kahane G, Savulescu J (2014) Autonomy and enhancement. Neuroethics 7(2):123–136
Schaefer GO, Savulescu J (2016) Procedural moral enhancement. Neuroethics; published, Online First, 20 Apr 2016
Shniderman AB, Solberg LB (2015) Cosmetic psychopharmacology for prisoners: reducing crime and recidivism through cognitive intervention. Neuroethics 8(3):315–326
Smith JA, trans. (1984) Aristotle, On the soul. In: Jonathan B (ed) Complete works of Aristotle, vol 1. Princeton University Press, Princeton, pp 641–692
Sorabji R (1980) Aristotle on the role of intellect in virtue. In: Rorty AO (ed) Essays on Aristotle's ethics. University of California Press, Berkeley, pp 201–219
Tappolet C (2010) Emotion, motivation, and action: the case of fear. In: Goldie P (ed) The Oxford handbook of philosophy of emotion. Oxford University Press, Oxford, pp 325–345
Urmson JO (1980) Aristotle's doctrine of the mean. In: Rorty AO (ed) Essays on Aristotle's ethics. University of California Press, Berkeley, pp 157–170
White NP, trans. (1983) Epictetus, Encheiridion. Hackett, Indianapolis
Zohny H (2015) The myth of cognitive enhancement drugs. Neuroethics 8(3):257–269

The Decision to Know: Pregnancy and Epistemic Harm

Kirsten M. Kringle-Baer

> *Knowledge seeking processes and its product are inextricably linked. Knowledge is a human creation and can only be as good as the efforts that go into attaining it.*
>
> —*Lorraine Code*

There is a sense in which the solution to the question 'Should we pursue knowledge of *x*?' might strike us as an easy one. Knowledge has been, and continues to be, understood as both intrinsically and extrinsically valuable, not just in the context of philosophy, but likewise in the context of our everyday lives. In what follows, I will challenge the inclination to assert a positive response to the aforementioned question without careful consideration of what 'knowledge of *x*' amounts to in the context of our deeply relevant, though often ignored, subjective life-circumstances. I will argue that, in some cases (and in one fully fleshed-out example in particular) pursuing knowledge is not always advisable. In cases like these, consideration of what I call 'epistemic harms' will be paramount in a decision of whether or not to pursue certain kinds of knowledge, and, additionally, that we can begin to develop a notion of wisdom as an epistemic virtue which will help guide us in these considerations.

With attention to both the subjective features of the knower and the object of knowledge which, according to Alcoff, are only available to us through a thick account of the circumstances, we shall see that there exist instances where we are better off not knowing, and, furthermore, that we require the epistemic virtue of wisdom in discerning when it is appropriate to pursue knowledge and when doing so may cause undue harm (2000).

In the narrative that follows, I recount my own experience with a pregnancy that was misidentified as high-risk. I hope that in sharing this account it will become evident that considerations of the circumstances of both the subject and the object of knowledge can impact whether it would be wise to pursue such information.

K.M. Kringle-Baer (✉)
Cincinnati, OH, USA
e-mail: Kirstenmk@gmail.com

L. Campo-Engelstein, P. Burcher (eds.), *Reproductive Ethics*,
DOI 10.1007/978-3-319-52630-0_4

Our case begins with a procedure that has become a greatly anticipated event in most women's pregnancies—the 20-week ultrasound. As I watched the screen breathlessly during the exam, 20 weeks into my first pregnancy, counting every wiggle, trying to discern the different body parts recognizable to me, the narration from the ultrasound technician slowly tapered, until finally the room was silent with the exception of deafening mouse clicks freezing and unfreezing images on the screen. I looked from the wand on my belly, to the screen, to the technician, and back again, reading appropriate worry from each. Finally, the technician excused herself to summon the doctor. The doctor, one I had not yet met, took his position at the machine and ran the wand over and along my swollen abdomen, without a single word. He shut off the machine and began, "well, here is what concerns us …."

This *ordinary* level-three ultrasound had revealed three markers of Down syndrome. The markers detected included a thick nuchal fold, an echogenic bowel, and choroid plexus cysts, all of which were conditions that, up until that point, I had never heard of. I was told that, alone, these markers didn't indicate any problem with my baby, but together, they increased the likelihood of having a baby with Down syndrome quite dramatically. Because I was so young, I had not elected to do any of the screening tests for genetic defects earlier in the pregnancy, but now I was urged to have my blood drawn in order to discern the likelihood of Down syndrome from yet another measure. I hastily agreed. Everything that day happened so quickly. In less than half an hour I went from a hushed, expectant anticipation in the waiting room to unrestrained sobbing in the exam room. I quickly submitted to taking tests I had actively chosen not to take just weeks prior (decisions made mainly based on how low risk this pregnancy *should* have been, and had been understood to be). Before I knew it I was in the car on the way home trying to decide what to tell my friends and family—whom to tell, even.

The 20-week ultrasound has become an assumed, if not required, diagnostic test in pregnancy. It is used to detect structural abnormalities of the fetus, the position of the placenta, and a myriad of other details about the pregnancy that are not discernable from the exterior of a woman's womb. They have been, for the most part, embraced by women as a chance to have a visual manifestation of their fetus, even in uncomplicated pregnancies (they are even performed commercially for this purpose!).

They are performed at 20 weeks—late enough so that measures of fetal growth and placental position are at least moderately relevant for projections of how the remainder of the pregnancy and labor will go—and early enough that, if abnormalities are detected, something can still be done about them. Here, the expression 'to do something' means to do further testing on the mother and the fetus, and, in some cases, to actually perform surgery on the fetus, to prepare for surgery after birth, to prepare the family for dealing with whatever abnormality has been detected, or to terminate the pregnancy. Such ultrasounds, however, are rarely pitched to women in this way—that is, if they are pitched at all.

There are some general worries about the use of ultrasound technology in pregnancy which are worthy of mention here. Indeed, they play a central role in how doctors can be assumed to become the best knowers and protectors of the fetus during pregnancy, and how mothers can become irrelevant or even an interference. These worries concern what the use of ultrasound does to our ideas, as a society, about the woman and the fetus, as well as what it does to the budding relationship between mother and baby. Ultrasounds turn the mother into something to be looked *through* in order to access information about the fetus. Ultrasound technology works by making the mother invisible in order to come to know things about the fetus. This fact subsequently results in an understanding of the fetus as extractable from, or independent of, the mother (Rothman 1993; Hilden 1996). Depicting fetuses themselves as somehow isolatable from the context in or conditions under which they live is understandably troubling and not only influences the relationship a mother has to her unborn child, but likewise has affected the way onlookers, whether family or society as a whole, come to understand pregnancy in general.

This worry should remind us of the urgings of Alcoff and Code: To take into account the situatedness of the knower and the object of knowledge. In this case, the situatedness of the object of knowledge is absolutely unique. According to Hilden, ignoring the location of the fetus, the status of being within another human being, leads to misinformed perceptions about the status of the fetus in society (1996). As a result of this kind of thinking, women can begin to consider the experience of seeing their fetus on the screen as somehow more enlightening and more meaningful than the *ordinary* experiences of pregnancy, the experiences available only *to her*—experiences which produce knowledge others can acquire only *from* her. So, here it is the relevant features of the object of knowledge which, when ignored, misguide us in the kinds of beliefs we develop about pregnancy.

In her essay, "Taking Subjectivity into Account," Lorraine Code argues that relevant factors concerning a subject's situatedness, such as race, class, or gender, ought to be considered in any investigation of what the subject can or cannot rightly be said to know. She writes, "conditions that hold for any knower, regardless of her or his identity, interests, and circumstances, in other words of her or his subjectivity—could conceivably be discovered only for a narrow range of artificially isolated and purified empirical knowledge claims, which might be paradigmatic by fiat, but are unlikely so 'in fact'" (Code 1996: 191). She argues for what she calls an "epistemology of everyday lives" and urges that we turn away from artificially sterilized models of knowledge that steer clear of what she takes to be the highly relevant circumstances of located knowers (Code 1996: 192).

Thus, she begins her own account of epistemology with a particular focus on the social dimension. This position is launched via an assault on traditional "*s* knows that *p*" epistemologies for their (often implicit) assumption that '*s*' and '*p*' are merely place holders which can be replaced by any *s* or any *p*, and maintain efficacy. When you begin to scrutinize the kinds of things you substitute for *s* and *p*, however, such a schema loses its capacity to represent many (most) natural knowledge claims. According to her interpretation of such approaches to representing knowledge claims, it follows that: "If one cannot transcend subjectivity and particularities of its 'locations,' then there is no knowledge worth analyzing" (192). Only a narrow subset of human knowledge claims, according to Code, can be isolated and analyzed by an epistemology which does not attend to the subjectivity of the knower.

It is Code's contention that, "the ideal objectivity of the universal knower is neither possible nor desirable, a realistic commitment to achieving empirical adequacy that engages in situated analyses of the subjectivities of both the knower and (where appropriate) the known is both desirable and possible … Objectivity requires taking subjectivity into account" (206). So, for an inquiry to be proper, it needs to consider the subjectivity of what is known and who is knowing it.

Here there is a real worry about potentially undermining the exclusive access a woman has to the pregnancy—as the sole person actually experiencing the *being pregnant*. There are many things that are available only to the woman, or via a direct examination of or conversation with the woman; for example, the woman typically feels the activity or movement of the fetus and can both describe and, if necessary "log" this information for the provider. How engaged the fetus is in (how far its head has dropped into) the pelvis can be determined externally as well. When the woman or the doctor encounter uncertainty, ultrasound could be turned to, rather, as a second option. This raises the question of what women can rightly be said to know about their own pregnancy. Again, I do not postulate a mysterious "sixth sense" that pregnant women have access too. What they do have access to it the experience of being pregnant with that child, and for this reason their exclusive access brings something to the table.

In weighing the issues that arise from the use of ultrasound technology in pregnancy, Barbara Katz Rothman writes,

> The doctor sits between mother and fetus. He turns away from the mother to examine the baby. Even the heartbeat is heard over a speaker removed from the woman's body. The technology that makes the baby/fetus more "visible" renders the woman *invisible* … The direct relationship to the baby within them, the fetus as part of their bodies, is superceded by the relationship with the fetus on the screen. The television image becomes more real than the fetus within; it is that image to which they "bond"; it is that image they hold in their minds as they feel their babies move. ("Tentative Pregnancy" 113)

The way in which an ultrasound is traditionally interpreted as a bonding experience is quite baffling when viewed in this light. In a sense, the child has never been further away from its mother than it is at the moment of an ultrasound!

In addition to what we have thus far encountered, there are worries raised by ultrasound use in pregnancy that are likewise relevant to this investigation. For example, ultrasounds have become so commonplace they are virtually unquestioned by mothers and health care personnel. Rarely do women elect *not* to have ultrasounds, and when they do, they often face opposition from their doctors or midwives. In all of the pregnancies and labors that turn out to be uncomplicated, these ultrasounds are unnecessary. Of course, it is impossible to know for certain that an ultrasound examination *will be* unnecessary beforehand; however, the decision of whether or not to perform one should always be left up to the mother, in the form of a genuine decision, not merely in having the (unknown) right to refuse the procedure.

Giving the pregnant woman this choice validates the authority she has to make decisions about her own body and her own pregnancy and allows her the opportunity to consider how valuable information about the fetus available through ultrasound is *to her*. Denying her the opportunity to make an overt decision about whether to have an ultrasound unduly privileges the kind of knowledge available through this technology, and, likewise, the doctor's desires, leanings, and interests over the mother's. Here it is attention to the subjective features of the *subject*, rather than the object of knowledge that is relevant. For some women, this information may be highly valuable; for others, not so, and for most, a chance to pause and pose this question to themselves will uncover that they hadn't yet given it adequate attention.

A further concern which urges us to push along in our narrative account is that ultrasound technology, though highly advanced, is *not* always a diagnostic tool; it, in this case and many others like it, is merely investigative. When an ultrasound reveals something unusual, the only course of action indicated is to *do more*. This point will be returned to later on.

Such was the case in my experience. Because the ultrasound indicated a *likelihood* of Down syndrome, the next step was to determine whether or not the baby was indeed affected. The only way to determine this with any degree of precision was via amniocentesis. After scheduling both an amniocentesis and a genetic counseling session (although my partner was not the genetic parent of the child), I came home to a freshly painted nursery, a box containing a crib intended to be built that weekend, even a significantly marked up book of baby names. I decided not to talk names. I decided not to put together the crib. I decided to tell people about the amniocentesis, and, if the test came back positive for Down syndrome *and* I decided to terminate the pregnancy, that I would tell my friends and family that I miscarried as a result of the amniocentesis. My (ironically) prolife partner suddenly started saying things like "I will support you if you decide to have an abortion," and "we can do this again—if we have to." I was horrified with her change in stance. Rothman writes, "seeking and waiting for information changes the pre-information stage of pregnancy, creates what I think of as a 'tentative pregnancy.' It incorporates the issue of abortion right into the route to motherhood and institutionalizes the conditionality in motherly love" ("Tentative Pregnancy" 1993: 85).

The talk of abortion, though, was set in motion prior to the comments my partner made to me. It started with the doctor who spoke with me right after my ultrasound. She came in to "explain" the findings of the doctor who oversaw the ultrasound. Terminating the pregnancy hadn't even crossed my mind until she indicated it as an option for dealing with a positive result from the amniocentesis, saying "we can take care of that for you too." "A one-stop shop," I thought to myself. My life was immediately put on hold by what I had learned (which, mind you, was not much of anything at all). All of my planning, all of my excitement became sour. I went from talking about "when the baby comes" to "if the baby comes." I went from collecting items for the baby to collecting receipts for the items we already had. I went from worrying about changing diapers to worrying about changing diapers for a lifetime. This expectant, paused state of being is precisely what Rothman means by the tentative pregnancy. Rothman believes that the practice of amniocentesis for genetic screening puts mothers and families into a detached and wary relationship with their developing fetus, a relationship characterized by a sense of unease and uncertainty.

This kind of relationship can cause a woman to experience some of the most exciting happenings of a pregnancy, such as the fetus's first felt kick, in a markedly different way than women not waiting for results from an amniocentesis to come in ("Tentative Pregnancy" 1993: 85).

The decision to have an amniocentesis and the related decision of whether or not to terminate the pregnancy if the results were positive collapsed into one for me. I barely thought about the amniocentesis. I barely thought about turning down an invasive, risky, and painful procedure, even after my blood tests returned indicating a much lower chance of any problems with the baby than that indicated by the ultrasound results. There seemed to be, however, no question of whether or not to proceed in collecting information. I needed *confirmation.* I don't know where this need came from, as I certainly hadn't felt it a week prior! Similarly, there was seemingly no question for the doctors or the genetic counselors I spoke with. I would have the test done.

The urgent and obvious nature of the affirmative answer to the first question of whether or not to have further testing done is indicative of the way in which a compulsion toward knowledge has influenced the medical management of pregnancy. The question of the value of knowing versus not knowing information about a pregnancy that is available through medical means is, in many ways, assumed to be answered by the medical establishment, and, predictably, answered in favor of the value of knowledge. This is a central issue to our discussion, and we will return to it toward the end of this chapter.

When I finally began to separate the two questions, one about amniocentesis and the other about termination, and turn to a consideration of the latter, it seemed that to others that decision, too, was an obvious one. To my partner, to my sisters, even to my own mother, the event of a positive result was the same as an affirmative answer to the question of whether or not to terminate the pregnancy. I had (excitedly) considered the fetus to be "my baby" up to that point. Suddenly, I was back to talking about "the pregnancy" instead of the baby—the kind of talk that characterized discussions before conception. I began to portray my experiences in light of this new way of talking, this new information—or lack thereof. Rothman writes, "The problem, or one of the problems, with the technology of amniocentesis and selective abortion is what it does to us, to mothers and to fathers and to families. It sets up a contradiction in definitions. It asks women to accept their pregnancies and their babies, to take care of the babies within them, and yet be willing to abort them" ("Tentative Pregnancy" 1993: 6).

For me, the appointments for genetic counseling and the amniocentesis coincided. I met with the genetic counselor, who took a family history (again, Down syndrome *is rarely* hereditary) and advised me to have the amniocentesis. Ultrasound is used while performing amniocentesis so the doctor can discern the best place to insert the needle and draw fluid with respect to the position of the baby and the placenta. The doctor chose, in our case, to push the needle directly through the placenta to draw fluid. This decision made the procedure more difficult and riskier. During the procedure, I watched the baby squirm and wriggle on the television screen. I watched the needle penetrate her sanctuary; I jokingly postulated that she disliked this. They finished the procedure by switching the machine to its three-dimensional mode and printing a picture of the baby. They told me it was a girl. Following the amniocentesis, the doctor recommended that we schedule a subsequent ultrasound to determine whether or not the choroid plexus cysts had resolved themselves. These, I was told, could pose problems in their own right. Several days later, I received a phone call from the genetic counselor. The baby did not have Down syndrome, nor any of the other genetic abnormalities detectable from the tests involving amniotic fluid. That evening, I finally chose a name for my baby girl.

At 6 months gestation, I came back in for the ultrasound to determine whether the choroid plexus cysts had resolved themselves. They had. The ultrasound technician, however, light-heartedly mentioned that the placenta looked "sparkly," a term she apparently coined and used to describe the appearance of increasing calcification of the placenta on the ultrasound screen. The placenta, normally depicted as fairly dark on the ultrasound screen, was adorned with a coat of little white flecks, indicative of a prematurely aging placenta. A placenta generally ages prematurely due to environmental factors, such as exposure to cigarette smoke—not a concern in my pregnancy, so I was not terribly worried at that point. At my subsequent appointment with the obstetrician (the same obstetrician who introduced the language of termination several weeks prior) the tone changed. We were urged to schedule weekly biophysical profiles and told that it was likely that they would need to deliver the baby early, since the aging placenta was not going to sustain her for the normal duration of pregnancy.

My pregnancy had so swiftly turned from something joyful and exciting to something anxiety-ridden and disheartening. Every week, for 2 months, I rode the two busses to get to the doctor's office, careful not to exert myself too much for fear of depriving the baby of oxygen, careful not to slip and fall on the December ice, to lie down on the table in the ultrasound exam room and watch my baby kick and squirm, swallow and expel the fluid that surrounded her. Every week the placenta, to them, appeared worse and worse. Every week it took a little longer for her to achieve all of the tasks she was meant to accomplish in the allotted 30-minute period (a developing fetus' movements tend to slow naturally as they grow in their confined space). Every week I left feeling more anxious, more fearful, that something would happen to her without my knowing it. I wanted to be able to constantly watch her on the screen. I wanted assurance that she wouldn't slip away from me in the days between my visits. Eventually, I got just that. The doctors decided that weekly exams were not sufficient, and asked me to begin coming in twice a week. The only time I felt reassured that she was indeed alive and well is when they told me so. Every time I felt her squirming or kicking I worried it was too frequent or too far between, as both, I was told, were indicative of a problem.

My body had turned into a treacherous place for a baby. It was as if while I was alone with her she was constantly in grave danger. The mere status of being inside of me put her unnecessarily at risk. She was only safe while being observed by the doctors, by individuals who could interpret her actions and translate them back to me in reassuring tones. I was told that passing a biophysical profile was "good for 48 hours," meaning that no harm would be expected in the 48 hours following one of these exams. This, of course, seemed absurd to me, but I was reassured again and again that it was so.

During the last few weeks of my pregnancy the weather turned very cold. I was happy for the excuse to curl up on the couch and closely monitor the baby's routine in my belly. I ventured out only for groceries (lots of them) and doctor's visits (lots of those, too). Finally the doctors decided that we should "come up with a plan" for her arrival. I was told that having one would relieve my anxiety, a complaint I was naive enough to mention when the doctor uncharacteristically asked me how *I* was doing. The plan they came up with was to do an amniocentesis to test for lung maturity at 36 weeks gestation and, in the event that the results were positive, to induce labor. The appointments were scheduled and, in a way, doing so did indeed alleviate some of my worries. I had an endpoint in sight. I only had to keep her alive for a few more weeks and, once she was out of me, she would be *safe in the doctor's arms*. I started worrying about lung maturity rates and induction procedures. I read up on premature babies, crossing potential hazards off my list as the pregnancy extended into 34 weeks, 35, then 36.

> But are women enemies of fetuses? Women, in fact, do not refuse such procedures nearly as often as they should … for most women, in the course of a wanted pregnancy, the fetus becomes real, precious, treasured. The overwhelming majority of women accept gratefully the cesarean sections their doctors offer—believing that it is best for the baby, even when the current data show quite clearly that probably three out of four cesarean sections in America are not necessary. (Rothman "Recreating Motherhood" 1989: 167)

The day of my scheduled amniocentesis and potential induction finally arrived. The ultrasound technician completed one final biophysical profile and printed out Carlin's last ultrasound photo, a barely visible profile of her gigantic foot. At this point, I had collected about 30 of them. My experiences of the moments leading up to her birth were so different from what I had imagined a mere 8 months previous. The doctor entered the room and began setting up for the amniocentesis. Because I had had one before, I started to worry about the pain of the procedure. Then I started to worry about the pain of childbirth.

Amniocentesis is considered to be more risky to a pregnancy earlier on. An amniocentesis is rarely done prior to 16 weeks because at any time sooner, a sufficient volume of amniotic fluid is not available to draw in order to run tests without serious risk to the pregnancy. The primary risk from amniocentesis is not, as you might expect, damage to the fetus but, instead, the induction of labor. The later into the pregnancy an amniocentesis is performed, the more likely it is that the fetus will survive if labor is triggered. At 36 weeks the risk of inducing labor was not as serious as it had been at 20 weeks. For this reason we moved on without hesitation.

Again, because of its prominent position, the doctor decided to go through the placenta to draw fluid. After several minutes of trying, the doctor removed the needle from my belly and I finally glanced back up at the screen. A rush of fluid was visible to me and, as the doctor hastily edged out of the room, I asked the ultrasound technician what I was seeing. "Blood," she answered, and through the doorway we heard the doctor's voice over the phone ordering a stat cesarean section. The needle had ruptured the placenta, and my blood was rushing into the amniotic sac.

The doctor returned to the room, along with a nurse and a wheelchair, hurriedly explaining what had happened and that the baby needed to be delivered right away.

As I was rushed through the hallways connecting the office building and the hospital, for the first time in several months I felt inexplicably tranquil. Here, the life of my baby was acutely at risk, by the hand of my own physician, and, instead of feelings of fear, I was entertaining a sense of relief. I knew that in a matter of moments, she would be safely in the world (as if she weren't somehow already)—no longer in a constant state of peril.

Because of the damage done to the placenta, there was no hope of natural birth, nor even induced labor. Instead, preparations were made for an emergency cesarean delivery. At this point a cesarean section *was* necessary because the uterus had indeed *become* an unsafe place for the baby. (Notice, it hadn't always been unsafe—a belief I was agonizing under for the last 2 months). The uterus was unsafe, however, not because it was in labor, but because it was quickly filling with blood.

I had read and heard about the procedure but was completely unprepared for what came next. After I donned the smock and endured a humiliating shave by my nurse, I found myself accompanied by just one unfamiliar nurse. I walked into the operating room, trembling. The room was bright, windowless, and cold. I sat on the bed and leaned forward so they could administer the anesthesia into my spine. As I started to become numb, I was situated into a supine position, with the lower half of my body completely exposed to everyone in the room but myself—a sheet was erected to further separate me from what was going on, to maintain a sterile field (my bottom half being more sterile than my top, of course). My arms were strapped down and I could feel nothing but cold—the cold of the air on my skin, the cold of the anesthesia working its way up my waist, the cold of the saline and anxiety medication winding its way through my veins. I wondered if the baby was cold too. Adrienne Rich writes,

> "but women are now asking what psychic effect a state of semihelplessness has on a healthy mother, awake during the birth, yet prevented from participating actively in delivery. No more devastating image could be invented for the bondage of woman: sheeted, supine, drugged, her wrists strapped down and her legs in stirrups, at the very moment when she is bringing new life into the world." (1976: 170–171)

They started cutting. I couldn't feel the cuts, but I could feel the pressure. I could tell they were doing *something* to me. I could tell they were doing things to me I would not be okay with if I were allowed to bear witness. I begged them to stop the surgery and wait until my partner was there, but they assured me she would be there in time for the delivery. I watched the clock; I prayed that they would slow down. I started to cry. Finally, the door opened and they allowed her in. They immediately hijacked her attention and began explaining what they were doing to me. I honestly do not recall if they had been explaining it to me all along and I just hadn't listened, but regardless, they were no longer talking to me. I told my partner I was scared and the anesthesiologist overheard. He said he would give me something to calm me down. They delivered the baby and, after weighing, washing, and drying her, they handed her to my partner, who brought her over to show me. When they took her away to be monitored, they called my partner to the other side of the sheet and showed her my placenta. The doctor said, almost triumphantly, "it looks good—except for the abruption!"

In another depiction of delivery, this time of a vaginal delivery, I find echoes of these same horrifying descriptions of women no longer in control of what is happening to their bodies:

> certainly a woman who was unconscious, semistupefied, amnesiac, or simply numb from the waist down cannot have experienced giving birth as an accomplishment, something over which she had no control. But what of the woman who is encouraged in childbirth preparation classes to see herself as a member of a "team" delivering her baby? Though she may help and watch in a mirror, she is not the primary actor. Positioning and draping her in such a way that she cannot directly see the birth, not allowing her to touch her genitals or the forthcoming baby, tells the mother that the birth is something that is happening to her or being done to her, not something she herself is doing. The birth is managed, conducted, by the other members of the team, those who are telling her what to do, and physically manipulating her and her baby. (Rothman "Giving Birth" 1982: 177)

There was something fitting in the fact that Carlin's birth had little or nothing to do with me. I was neither an active participant nor particularly informed about what was going on. From the moment that I had my 20-week ultrasound until the moment she and I were released from the hospital, Carlin was cared for and managed by someone other than me. Her first 36 hours in this world were characterized by the dim lights and constant beeping of the NICU, difficult IV placements (which are nothing like IV placements for adults, with the most prominent veins in infants being those in their heads), and one delirious mom, heavily medicated and recovering from major surgery, peering over her plastic container. Diapers were changed and weighed by nurses; food was administered intravenously. She was finally released to "my care" only to be hovered over and eventually removed from my hospital room on account of "low body temperature." Later I was told by our pediatrician that the best remedy for low body temperature in a newborn is skin-to-skin contact with the mother, not the warming lamps they lay infants under like french-fries. Breastfeeding sessions were observed and critiqued by lactation consultants.

When we were finally released from the hospital, 24 hours passed in a dreamlike haze before Carlin was readmitted to the hospital on account of complications stemming from her premature birth and the ingestion of my blood as a result of the ruptured placenta. Even after being taken from me and delivered to supposed safety, antibodies from my blood that had entered her blood stream were attacking her and preventing her from thriving. I was endangering her from a distance, even.

She spent the next month in a world she shouldn't yet have been in, losing weight, suffering from jaundice, and having her heels poked, with a medicated mom recovering from a surgery she needn't have been subjected to. Carlin's premature delivery and difficult first month in the world were the direct result of an over-managed pregnancy—a pregnancy characterized by fear and anxiety, observation and intervention, and the undermining of the confidence of a mother and the developing relationship between mother and child.

While it is acknowledged that hindsight is always 20/20, it should be pointed out that the only test that would have revealed a genuine problem with Carlin was the botched 36-week amniocentesis to test for lung maturity. No amniotic fluid was successfully drawn during the procedure, but the placental abruption caused by it necessitated an early delivery regardless of whether or not her tiny lungs were prepared to take in air. And this abruption was the only problem they noted with the placenta following delivery. That is, there was no evidence of premature placental aging, or any other condition that could have caused harm to the baby. What did end up causing a problem for her was the presence of my blood in the amniotic fluid, blood that was introduced as a result of the unsuccessful amniocentesis.

What really struck me as I gazed down at my average sized infant in the NICU was that she was the product of a completely healthy pregnancy. The only reasons I could come up with to explain the tubes and monitors coming off her had nothing to do with me. Acknowledging this fact, standing in stark contrast to everything I had been told and everything I felt up to that point, immediately relieved the cognitive dissonance built up inside me by wanting so badly to take care of my daughter and yet knowing that I was putting her in danger just by having her inside of me.

The use of indiscriminate investigative procedures such as 20-week ultrasounds can result in the overdiagnosis of problems with the fetus. This, of course, is particularly worrisome in the case of patients who are not at risk. Once the alarm has been sounded, mothers and doctors alike are compelled to continue down the path of diagnostic screening and intervention. It would be difficult, if not impossible, to turn away from the risk of Down syndrome, or, even more so, a risk of the baby's not thriving due to a prematurely aging placenta. Once the problem has been identified, whether correctly or incorrectly, a path for action has been laid down for mother and physician alike.

Even if we bracket the possibility for the kind of harm *set in motion* by these information-seeking techniques suggested above, the question arises as to whether harm can come from the mere information seeking in the first place. In other words, if we grant the possibility that procedures like screening for abnormalities of the fetus with ultrasound were 100% accurate in predicting problems, would there still be a concern about the use of ultrasound, that is, the asking of these kinds of questions, in the first place? The answer, I think, is clearly 'yes.' There is, in the case of pregnancy, something troubling in the very *asking* of these questions. A certain kind of information or knowledge about the fetus has come to be valued over another.

How is it that we have come to be asking the questions that we ask about pregnancy? Why do we need to know prior to delivery if the baby is a boy or a girl, healthy or unhealthy, upside-down or right-side-up? In the case of some of these questions, the answer is obvious: Because we can *do* something about it. For others, the answer is not so obvious. For others, the answer might be something more like, because we *can know*. To me, it is not at all clear why knowing, in this case, is any better than not knowing and, moreover, why knowing in this privileged, empirical, and scientific way is better. There is a certain kind of epistemic imperialism at play here. This epistemic imperialism sets out what the important questions are and sets out the ways in which we are to go about answering those questions. The kinds of questions we are asking and the kinds of answers we are getting are driven by, created by, the kinds of technology we have—not, instead, by any identifiable *need* we have for this particular kind of information. This sets up a false hierarchy of knowledge—privileging the information that is technologically available, making it seem more important when, really, it is only more available, and available to more professionalized, and presumably more reliable, knowers. The influence of social values, or even merely the interests of the medical and scientific community, are, in line with the suggestion of Jaggar, establishing the questions we are asking, recommending the routes we take in answering them and, clearly, offering up the answers. Jaggar suggests that we are being unrealistic in supposing that knowledge-seeking can be free from the influence of social values. She writes, "these values are implicit in the identification of the problems that are considered worthy of investigation, in the selection of hypotheses that are considered worthy of testing, and in the solutions to the problems that are considered worthy of acceptance" (1996: 176). The questions themselves in the case explored here and many others like it are being generated based not on a *need* to know, but rather on an *ability* to know.

In an alternative reading of Shakespeare's *Othello*, Naomi Scheman gives an account of what she takes to be the *real* harm committed by Iago (370). Her interpretation suggests that it is not that Iago convinces Othello that Desdemona has been unfaithful but instead that Iago changes the terms, or the nature of the evidence, required to establish feelings of mutual trust in their relationship. She writes:

> Iago's skillful manipulation of the appearances (he doesn't exactly *lie*) is not a perversion of scientific reasoning, but, in its power to seduce Othello, a demonstration both of the incapacity of such reason to comprehend aspects of the world that lie beyond it and the defenseless inability of that world to provide a logical, rational proof of its own reality. It needs—demands—no proof, but pressed to give one, it will inevitably fail. (1998: 370)

This can be seen as a move similar to the one being made by medical professionals in the case of the relationship between mother and unborn child. What was once a relationship built on the emotional and intellectual insights of the mother (and the other women surrounding and attending to her during the pregnancy and labor) and the physical sensations of the mother alone is now a relationship forced to fit into an unfamiliar mold, subjected to novel questions as well as procedures for answering those questions which are responsible for changing the character of their relationship.

We ought to ask ourselves who Iago is to set out the questions at issue in the relationship between Othello and Desdemona. We ought to ask ourselves who the doctor is to set out the questions at issue in the relationship between mother and child. It is time, then, to establish *our own* questions, to determine for *ourselves* reliable methods of answering them (methods that might well call on medical expertise), and to provide some tentative answers. The question proposed here, again, is whether knowledge is always more valuable than ignorance. The reasoning behind the response I advocate involves a recognition of some potential 'epistemic harms.' Epistemic harms can be understood as very much like physical harms: They are the intellectual harms suffered by the knower and the known alike resulting from certain types of knowledge gaining procedures and, in some cases, from the knowledge gained itself. We have encountered several of these already, including the impact of ultrasound use on the perceptions of the mother regarding her baby and the thrust of probabilistic knowledge of risk to pursue more and more evidence to engender worry and fear. Let us look at some more.

According to Amy Hilden, pregnancy should be about developing a relationship with the fetus, about experiencing certain changes in your body and your life that are characteristically female (100). Pregnancy, moreover, should be about a growing sense of confidence in yourself as a mother. When medical intervention procedures are allowed to escalate out of control, it changes the whole tone of pregnancy. It can turn pregnancy into a medical crisis. It punctuates the felt flips and kicks of the fetus with fearful rather than joyful anticipation. The escalation of intervention in pregnancy undermines the epistemic authority and control that the woman has over her pregnancy. When an individual's attempt to make meaning for herself is thwarted by a society's or an institution's incompatible understanding of an issue, Fricker deems it a case of hermeneutical injustice (2007: 155). She writes:

> When you find yourself in a situation in which you seem to be the only one to feel the dissonance between received understanding and your own intimidated sense of a given experience, it tends to knock your faith in your own ability to make sense of the world … [it] stem[s] most basically from the subject's loss of epistemic confidence. The various ways in which loss of epistemic confidence might hinder one's epistemic career are … that it can cause literal loss of knowledge, that it may prevent from one gaining new knowledge, and more generally, that it is likely to stop one gaining certain important epistemic virtues, such as intellectual courage. (163)

Two different yet intertwined claims that women have to knowledge about pregnancy, I believe, need to be disentangled here. On the one hand, the historical practice of midwifery and the techniques, approaches, and insights passed down by women to women throughout human history which has now been replaced by the modern, medicalized approach to pregnancy we are more familiar with today do indeed suggest that there is some kind of special access, some practical insight women have to matters of managing pregnancy (Rich 1976: 149). Even if we bracket this possibility, there is yet another kind of knowledge that women have access to that is or can be challenged by technologically driven ways of knowing about pregnancy; women who are pregnant, or who have been pregnant, have what is called experiential knowledge of their pregnancy and of pregnancy in general. This kind of knowledge is not to be taken lightly and has the potential impact of altogether shifting the way an individual sees the world, and sees herself in the world (Shapiro 2009, 2010: 59). This kind of knowledge, however, is not meant to compete with or mimic the kind of knowledge available through the evidence-based, empirical information-seeking technologies used in pregnancy. It is of a different kind altogether. At the same time, this kind of knowledge is indeed threatened, rather than being helpfully supplemented, by these technologies. In fact, this kind of knowledge takes a back seat to the kind available to the physician by looking *through* the mother, whether it be by needle or by sound wave.

With these two distinct flavors of potential knowledge only women have about their pregnancies in mind, it is not the case that I am arguing that I, or any other woman, is in a position to know better than my doctors about the health and viability (or lack thereof) of my daughter *in utero*. What *is* being argued is that the preference for having the kind of knowledge available through medical diagnostics may be unfounded, and the quest for this kind of information disrupts the course and experience of pregnancy and its epistemic consequences. The privilege of and search for this kind of information about the pregnancy, however accurate or inaccurate the results, pushes aside the importance of the kind of knowledge a woman *can* have about her pregnancy, dramatically alters the way she experiences it, and depletes the sense of (and actual) control she has over how it unfolds.

With adequate weight given to the sorts of epistemic harms caused by the pursuit of certain kinds of knowledge, through certain means, I hope I have shown that the search for knowledge can sometimes be inadvisable. That is to say, sometimes, knowledge *isn't* more valuable than ignorance.

Searching for Solutions

With these concerns in mind, it becomes prudent in closing to ask ourselves what might have been done otherwise. In answering, a frequently quoted line from § 308 in Wittgenstein's *Philosophical Investigations* comes readily to mind: "the decisive movement in the conjuring trick has been made, and it was the very one that we thought quite innocent" (103). Although undeniably taken out of context, we might interpret this to mean that we should start asking questions sooner. It was at the 20-week ultrasound, a procedure I excitedly anticipated and in no way scrutinized, that the ball began rolling in this particular instance of escalation of intervention. The very asking of the kinds of questions answerable by medical technology in pregnancy can have, and has had, the impact of setting off a destructive chain of events—a sequence of questions, answers, and approaches, that have the effect of causing the kind of epistemic harm to the mother explored in these pages.

The discipline of virtue epistemology, however, gives us a framework in which we *are* positioned to do more. To remind ourselves: First, it made room for our novel inquiry in a discipline characterized by often rigid adherence to a pre-established set of questions deemed worthy of investigation. Second, it justified our turning to the subjective features of both the knower and the known in finding answers to our questions. Third, in carrying out such an investigation, it highlighted some salient outcomes, namely, the epistemic harms, which undoubtedly deserved a closer look, and, when given one, provided an answer to the question at hand. Finally, then, virtue epistemology will help us turn our results into something pragmatic, an aim which many virtue epistemologists consider the proper end of epistemology in general.

Valerie Tiberius proposes we define wisdom as "the virtue that allows us to make choices and act in such a way that we can reasonably expect to achieve a satisfactory review of our own conduct" (215). To pursue knowledge wisely, then, amounts to pursuing knowledge in a way that allows for a "satisfactory review" of ourselves. Such a satisfactory view, I propose, necessarily takes into consideration the various epistemic harms caused by any potential pursuit of knowledge.

The epistemic virtue of wisdom, when turned in the direction of questions of whether or not to pursue a given line of inquiry, can help us sort through the potential epistemic harms, can help us attend to the relevant features of the subjects and objects of knowledge, as well as the community in which they are situated, and, in turn, help us provide an adequate response. If we are to act in accordance with wisdom, then, we are not to assume that the answer to the question of whether it is valuable to know *x* is either an unqualified 'yes' *or* 'no.' Instead, if we are indeed to achieve a satisfactory review of ourselves, the question of the value of knowledge will vary with respect to the relevant features of the knower and the known, and the epistemic harms likely to be produced by such an inquiry for both, as well as for the community in which they exist.

The account I have given of wisdom as an epistemic virtue undeniably deserves a more thorough exposition. In developing one, it is my belief that we need to look not just to the work of philosophers, but to the voices of researchers in psychology as well, for example, how we reason when faced with probabilistic information, in particular, in medical decision-making. According to Reyna and Brainerd, many people struggle with numeracy and, in particular, with probabilities (2008: 89). They identify several common mistakes we make in judging probabilities and risks and remind us that these are abilities that are required for informed medical decision-making. Although they fall short of providing a way for us to avoid these mistakes in reasoning, merely having been made aware of them has caused me to look more carefully at numerical information, especially when representing medical risks. Sedlmeier (1999) presents research into several training regimens which have been established to improve statistical reasoning, some more successful than others. While I don't advocate a battery of training regimens in statistical reasoning as required for developing the virtue of wisdom, I do believe there are important insights to be gained from what psychologists have determined has worked and what hasn't in improving our ability to reason well with statistical information.

While our account of wisdom is not a complete one, I believe that it can be properly filled out by attending to research into decision-making (medical decision-making in particular) as well as looking to other virtue-epistemologists for useful elements in their own accounts of wisdom. Elsewhere I explore these avenues more thoroughly and have found it to be a worthwhile endeavor.

It has, I hope, been shown that any response to the query of whether knowledge is more valuable than ignorance must, if in accordance with wisdom, attend to the subjective features relevant to the inquiry (those which have been traditionally overlooked by epistemologists) as well as weigh the potential for resulting epistemic harm. These claims are, at minimum, surprising, and, at best, innovative, given the leanings of the technologically-driven and information-oriented society in which we live. In knowledge seeking practices ranging from the ordinary Google search by an individual to the carefully developed investigations of highly regarded research institutions, we often attend to the potential *benefit* of gained information, but rarely regard the costs that can be incurred. We are too inclined to think that considerations of potential harm enter in only when we ask what we might do in light of knowledge we have obtained. What I hope to have shown is that pursuit of knowledge itself can be harmful and that we need to think of that pursuit as the "initial move in the conjuring trick," one that escapes notice but that can be crucial to the outcome. It is my contention, then, that the epistemic harms outlined in this paper be taken as seriously as physical harms, and, as such, consideration of them ought to precede investigation, when relevant. And where should we look for guidance in sorting through these considerations? To this, my answer is: The epistemic virtue of wisdom.

References

Alcoff LM (2000) On judging epistemic credibility: is social identity relevant? In: Zack N (ed) Women of color and philosophy. Blackwell Publishers, Malden

Code L (1996) Taking subjectivity into account. In: Garry A, Pearsall M (eds) Women, knowledge, and reality: explorations in feminist philosophy. Routledge, New York

Fricker M (2007) Epistemic injustice. Oxford University Press, Oxford

Hilden AB (1996) Pregnancy as a developing relationship: implications for the construction of fetal personhood. Dissertation, University of Minnesota

Jaggar A (1996) Love and knowledge: emotion in feminist epistemology. In: Garry A, Pearsall M (eds) Women, knowledge, and reality: explorations in feminist philosophy. Routledge, New York

Reyna V, Brainerd C (2008) Numeracy, ratio bias, and denominator neglect in judgment of risk and probability. Learn Individ Differ 18:89–107

Rich A (1976) Of woman born. W.W. Norton & Company, New York

Rothman BK (1982) Giving birth: alternatives in childbirth. Penguin Books, New York

Rothman BK (1989) Recreating motherhood. W.W. Norton & Company, New York

Rothman BK (1993) The tentative pregnancy: how amniocentesis changes the experience of motherhood. W.W. Norton & Company, New York

Scheman N (1998) Othello's doubt/Desdemona's death: the engendering of skepticism. In: Alcoff LM (ed) Epistemology: the big questions. Blackwell Publishing, Malden

Sedlmeier P (1999) Improving statistical reasoning: theoretical models and practical implications. Lawrence Erlbaum Associates, Mahwah

Shapiro D (2009) Experiential knowledge: the knowledge of "*What it's like*". Dissertation, University of Minnesota

Shapiro S (2010) Decision making under pressure. Futurist 44:42–44

Part II
Fertility Preservation

Egg Freezing and the Feminist Quest for Equality in the Workplace

Karey Harwood

Introduction

This paper discusses how non-medical egg freezing fits both symbolically and pragmatically into the feminist quest for equality in the workplace. I argue that egg freezing, on balance, works at cross purposes with ongoing efforts to make real structural changes in the American workplace. Before elaborating this argument, I first review background information about egg freezing and some of the ethical arguments that both support and critique women's use of it for non-medical or "social" reasons.

For the purposes of this paper, I do not discuss the use of egg freezing for medical reasons, such as freezing one's eggs prior to undergoing chemotherapy. I focus exclusively on egg freezing undertaken for the sake of "buying time"—time to find a partner, time to finish an education, time to advance in one's career, or any conceivable reason *other* than facing the crisis of an immediate medical need to preserve one's fertility against likely iatrogenic damage.

Current Trends in Non-medical Egg Freezing

A cycle of egg freezing begins like any cycle of in vitro fertilization (IVF), with the administration of hormones to stimulate a woman's ovaries into producing multiple eggs, followed by careful monitoring of the ovaries, and then finally the retrieval of mature eggs under anesthesia. After the eggs are extracted, rather than being combined with sperm as they would be in a typical IVF cycle intending to create

K. Harwood (✉)
Department of Philosophy and Religious Studies, North Carolina State University, Raleigh, NC, USA
e-mail: karey_harwood@ncsu.edu

L. Campo-Engelstein, P. Burcher (eds.), *Reproductive Ethics*,
DOI 10.1007/978-3-319-52630-0_5

Illustration A https://www.eggbanxx.com

embryos, the eggs are frozen—usually through a technique known as vitrification. The frozen eggs are then stored until such time as the woman wants to attempt fertilization and pregnancy.

In 2012, the American Society of Reproductive Medicine declared that egg freezing should no longer be considered experimental.[1] This was a significant announcement. Equally significant was the fact that the ASRM explicitly discouraged egg freezing "for the sole purpose of circumventing reproductive aging in healthy women."[2] Notwithstanding this note of caution, the commercial promotion of egg freezing for avoiding age-related infertility has proceeded apace ever since. Companies reach potential customers through websites and meet and greet cocktail hours, and advertise access to fertility specialists as well as special pricing for this generally out of pocket expense.

For example, EggBanxx, a relatively new startup in New York City, offers financing and connects women with doctors who specialize in egg freezing [Illustration A]. When I visited the EggBanxx website from my office computer in Raleigh, NC, it automatically located my nearest egg freezing provider and quoted me a price of $8500 for one cycle.[3] Extend Fertility, a company launched in 2004, similarly connects women with a network of fertility specialists and provides financing.[4] As of April 2016, Extend Fertility offered an introductory rate of $3600 per cycle, not including egg storage, claiming a 25–50% discount off their usual fees [Illustration B].[5]

By some estimates, the number of women seeking non-medical egg freezing has doubled every year since 2010.[6] However, egg freezing is still a very small subset of the assisted reproductive technology landscape. Out of the 190,773 cycles of Assisted Reproductive Technology (ART) performed in 2013 and tracked by the

Illustration B http://www.extendfertility.com

Centers for Disease Control and Prevention, only 1.4% were initiated for the purpose of thawing a frozen egg for fertilization. A larger proportion of ART cycles, 27,564 (14.4%), were undertaken with the goal of freezing or banking the resulting eggs or embryos [Illustration C].[7]

The CDC has been required by law since 1992 to report on the success rates of Assisted Reproductive Technology. This it does quite thoroughly, covering an estimated 98% of ART cycles in 2013. However, neither the ART cycles that made use of a frozen egg to attempt fertilization and pregnancy, nor the ART cycles that resulted in the banking of eggs or embryos, were included in clinic success rates reported by the CDC.[8]

As a benchmark for comparison, the success rate for fresh (non-frozen), non-donor ART cycles was 29% in 2013, with "success" here being defined as a live birth. Notably, almost 1 in 5 ART pregnancies did not result in a live birth.[9] Could success rates for IVF using a previously frozen egg be as high as 29%? Some say yes. Some claim even higher, including Dr. Nicole Noyes of the NYU Fertility Center.[10] The best available evidence suggests that pregnancy rates for fresh and frozen eggs are about the same.[11] Given the disputed nature of these statistics, perhaps the claim with the strongest evidence supporting it is this: the age of the woman at the time she freezes her eggs affects the subsequent success rate. Success rates appear to be significantly lower for women who freeze their eggs after the age of 38.[12]

Unfortunately, studies have shown that women do not consider egg freezing until their late 30s or older. The ideal age for egg freezing is reportedly 30–35, but these younger women tend not to consider egg freezing either because they believe they have plenty of time and/or they underestimate their natural fertility decline.[13]

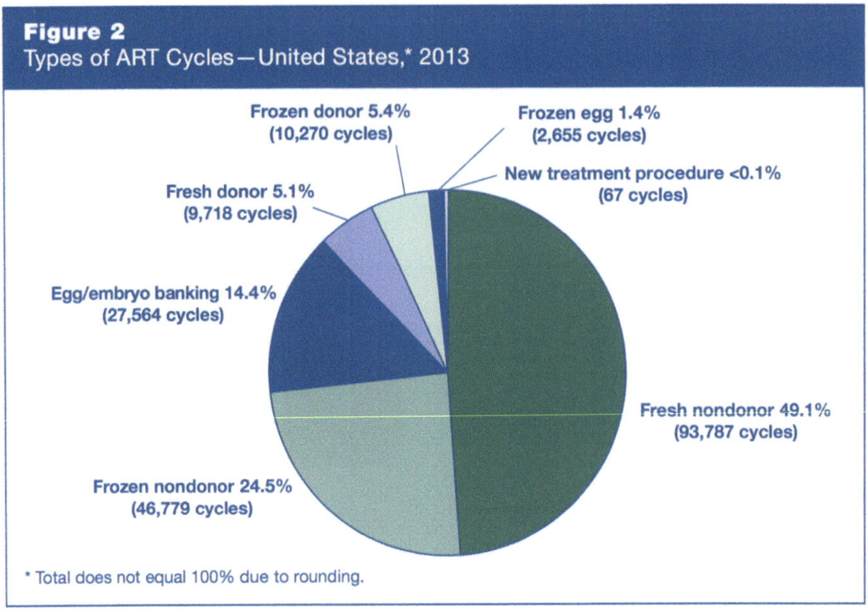

Illustration C 2013 National Summary Report, Centers for Disease Control and Prevention http://www.cdc.gov/art/pdf/2013-report/art_2013_national_summary_report.pdf

Current Trends in Delayed Childbearing and Women's Employment

EggBanxx and Extend Fertility may have a point that "Smart Women Freeze" and "Preserve Their Options." The biological reality is that women's fertility declines after age 35. The social reality is that there is a steady trend toward delayed childbearing across all subgroups of women.

In 2014, the average age of first-time mothers was 26.3, which represents an increase of more than 5 years from 1970, when the average age of first-time mothers was 21.4.[14] Another way to look at the data is to consider the proportion of first births that are occurring for women aged 35 and over: "From 1970 to 2006 the proportion of first births to women aged 35 years and over increased nearly eight times."[15] One out of 100 first births in 1970 were to women 35 and older. That proportion was 1 out of 12 in 2006.[16]

There are likely many causes for this trend toward older motherhood, which has arguably brought many positive benefits for women, including more time to complete an education and more opportunities to pursue careers. What is important for my purposes in this paper is simply to acknowledge that the use of egg freezing is taking place inside a social context of delayed childbearing and inside a wider conversation about women in the workplace. In her book, *Unfinished Business: Women,*

Men, Work, and Family (2015), Anne-Marie Slaughter summarizes some of the well known facts covered in sociological literature: the majority of women with school aged children work outside the home in the United States,[17] women lose significant lifetime earnings when they step out of the workforce to raise children,[18] the United States still lacks paid family leave and universal childcare, and men still do not contribute as much in "care work" or household labor as their wives do, even if they are both employed outside the home.[19] Also, young women and men first starting out in careers make comparable incomes—there is hardly a wage gap at the outset. And they are reportedly equally ambitious.[20] But women with children earn less than men with children, and the gendered inequality increases over time. As many feminist have argued, the structure of the workforce still assumes a fully committed worker who relies on the domestic management of an at-home partner, even though the economic reality of the twenty-first century is that many American families need two wage earners to achieve a standard of living previously possible with only one income.[21] Given this larger context, where does egg freezing fit into the feminist quest for equality in the workplace?

Goold and Savulescu's Defense of Egg Freezing as Reproductive Affirmative Action

In their 2009 *Bioethics* article, "In Favour of Freezing Eggs for Non-medical Reasons," Imogen Goold and Julian Savulescu argued that non-medical egg freezing can serve as a form of "reproductive affirmative action," a strategy for leveling the reproductive playing field for women that does not diminish larger efforts to create structural change.[22] According to these authors, women may freeze their eggs to elongate the time they have to pursue education and career, for example, and these private choices need not impede the "unfinished business" of reforming the workplace, such as advocacy for national paid maternity leave or a work culture that provides "off/on ramps" for women who interrupt their careers for childbearing.

The general benefits they cite for non-medical egg freezing are compelling at first blush: egg freezing reduces pressure on women by stopping or pausing the reproductive clock. It affords more time to find a partner, and more time to be emotionally and psychologically ready to be a parent. Egg freezing can even be a type of insurance, guarding against the unknown or not entirely controllable trajectory of one's life course.[23]

Goold and Savulescu also explicitly argue that egg freezing promotes gender equality: by freezing their eggs, women can participate equally with men at work. They write, "The 30s represent a crucial time in the careers of many women, and an interruption to bear children at that point can seriously prejudice a woman's chance of advancement."[24] The authors are realistic about social structures that disadvantage women, acknowledging that "full participation by women in the workforce is not a reality" in some societies.[25] But they do not abandon the possibility of social change. I quote them here at length:

> ...[I]t is true that many of the problems we cite stem from traditional employment models that are based on the employee being male. We should pro-actively seek to change this situation to ensure that women have the opportunity to pursue a career as they choose, rather than having to fit into a model designed without them in mind. But such a desire for change is not necessarily undermined by allowing access to technological advances that can remove some of the constraints women face in their employment. We can pursue various secondary strategies, including egg freezing, for improving women's employment situation.[26]

Goold and Savulescu conclude optimistically with the prediction that "egg freezing will probably not undermine efforts to promote change"[27] and that it might instead enable women to achieve positions of authority and influence. Once well situated, such successful women would be in a position to reform the workplace, they argue. "Thus, egg freezing can be viewed as kind of reproductive affirmative action: when discriminatory features of society are changed, it may no longer be necessary. But in the meantime, in our view, it empowers women."[28]

While I appreciate the pragmatism of Goold and Savulescu's approach—they accept that society is the way it is for now, and judge egg freezing to be a progressive technology that aids women—I find their argument ultimately unsatisfying for these reasons: (1) their argument depends upon a mindset of private consumer choice that is problematic, primarily because it creates a fiction of planning and control when the reality of people's lives suggests otherwise, but also because that fiction of planning and control fits too conveniently with the opportunity to profit from a procedure that carries health risks and a relatively low probability of success; and (2) shifting the burden to individual women to accommodate work structures is not preferable to changing work structures to accommodate women, primarily because deflecting responsibility for a needed structural change does little to create lasting justice. I will spend the remainder of paper explaining these reasons.

Rhetoric of Choice and Consumerism Is Problematic, and Succumbs to Profit-Drive

Egg freezing fits well with the narrative of choice and consumerism. Finding paying customers for egg freezing in fact depends upon the assumption that women approach their reproductive lives this way: as a subject of planning and foresight. Remember, "Smart Women Freeze." This is, allegedly, a liberating vision. It imagines women empowered by technology, privileged in their options, able to use their resources to stretch and mold what is biologically possible.[29] It imagines both a postmodern playground of self-creation[30] and a savvy consumer's gift for creative problem-solving.[31]

In reality, delayed childbearing is not necessarily a conscious choice and potential consumers of egg freezing often come late to the realization that they need it—notwithstanding marketers' attempts to convince them otherwise. A recent qualitative study in the UK found that delayed childbearing is rarely a conscious choice. The study reported that women's "timing of childbearing depended on a complex interplay of factors which were outside of their control, such as relationship, health, and fertility. Women felt that there was an element of chance that these factors would be in place at the time they desired."[32]

Feminist philosopher Angel Petropanagos argues that women should not be blamed for their "perpetual postponement" of pregnancy. She brings a feminist understanding of relational autonomy to her assessment: "Unlike traditional accounts of autonomy, feminists' accounts of autonomy require an explicit recognition that autonomy is both defined and pursued in a social context."[33] Given the context of sexist social structures, she argues, including professional norms, the "choice" to delay childbearing may not be as voluntary as it seems. "Until the sexist social structures that shape and confine women's reproductive choices change, many women may continue to find their lives unfolding in ways that result in delayed motherhood."[34]

However, recognizing that women's reproductive decision-making is socially constructed and constrained by many things can lead to divergent conclusions. Some, like Petropanagos, and Goold and Savulescu before her, see the compensatory value of egg freezing and support it. Others wonder more skeptically whether egg freezing may do more harm than good. The problem, according to this view, is that the option of egg freezing creates a pressure to use it.

As philosopher Michael Sandel once claimed in his discussion of hypothetical problems created by genetic engineering, offering the option to genetically engineer one's children could engender a sense of hyper-responsibility in parents for their offsprings' traits.[35] In an environment where intervening genetically, pre-birth, became the norm, it would no longer be a misfortune if one's child were short or nearsighted or hemophiliac; it would now be the parents' fault for not taking advantage of an available technology proactively.

Similarly, as pointed out when Apple and Facebook made their generous offers to provide egg freezing benefits to their employees, the option to freeze eggs might quickly become an obligation to freeze eggs, a way to demonstrate one's seriousness about one's career, or a way to avoid self-blame.[36] More generally, in an environment where taking steps to guard against age-related infertility became the norm, it would no longer be merely a misfortune if one ran out of time to have children; it would be the fault of women themselves for not taking advantage of an available technology proactively. In fact, there is already evidence from an empirical study of women's motivations that a major impetus for egg-freezing is the avoidance of self-blame. One woman stated: "I want to know I did everything I could and not blame myself later."[37]

Or, as another writer, Abby Rabinowitz, succinctly put it: "Egg freezing is the perfect regret machine."[38] Egg freezing is an example of the market not just responding to a consumer need, but amplifying and exploiting one. It mines a deep vein of culturally constructed expectations of women vis-à-vis motherhood.

I would add to this point an additional concern that as women buy into the fiction of planning and control offered by egg freezing, they also, not coincidentally, fuel the profits of companies like EggBanxx. This is not in itself a problem, nor is it necessarily exploitation. But given the low probability of success, especially for older women who freeze their eggs, this vulnerability should raise the bar for the kind of detailed information about risks and success rates that are made available to potential consumers of this technology.

Feminists have long embraced the language of choice for the obvious reason that it is fundamentally connected with reproductive autonomy and politically connected with the pro-choice movement. While empowering in some respects, the rhetoric of choice can be self-defeating if it narrows our understanding of reproductive autonomy to the realm of private consumer choice. The neoliberal free market mindset of consumption tends to obscure structural constraints on individuals' actual choices, inflating our sense of agency and diminishing the urgency of needed social change.

A Technological Fix for a Social Problem Works at Cross Purposes with Real Reform of the Workplace and Long-Term Demands of Justice

This point brings me to my final reason for questioning the helpfulness of egg freezing as a form of "reproductive affirmative action": I contend that shifting the burden to individual women to accommodate work structures is not preferable to changing work structures to accommodate women, because deflecting responsibility for a needed structural change does little to create lasting justice.

For one, egg freezing provides a relatively easy "out" for companies if they offer the service as an employee benefit. It is a way to look as if they are providing support for employees' procreative aspirations without having to the change anything about the culture of work that often necessitates women's deferral of childbearing in the first place.

Also, egg freezing, pragmatically speaking, is not more than a very small bandage compared with the systemic difficulties faced by women in the workplace and society generally.[39] It is a very privileged band-aid too, accessible only to those who can afford to pay out of pocket. Moreover, contrary to Goold and Savulescu's prediction that women who attain positions of power due to this form of reproductive affirmative action would eventually look to reform the workplace, there is no real evidence such women exist. Nor is there much evidence, according to Anne-Marie Slaughter, that women who achieve success within the framework of a patriarchal work structure demonstrate any inclination to change the structure that they themselves labored under and triumphed over.

Symbolically, I would argue egg freezing sends an out-of-step message. One might even say that the commercial promotion of egg freezing comes at an odd time given some other notable trends. As reported recently in *The New York Times Magazine* by Susan Dominus, some companies are beginning, finally, to make some significant strides in rethinking work-family balance, showing greater awareness of the costs of overwork, and a greater willingness to change.[40] Perhaps inspired by Millennials, who tend to question the fantasy and sanity of "having it all," there is a dawning realization that the structure of the workplace not only disadvantages women but devalues caregiving generally, and this ultimately, in the long run, harms the bottom line. Some businesses are beginning to realize that work-family "balance" needs an entirely new frame.

For example, a recent sociological study of a "Results Only Work Environment" (ROWE) found that giving employees control over the hours of their work and the location of their work, provided they get their work done, alleviated work-family conflicts and increased job satisfaction.[41] A recent op-ed in *The New York Times* even claimed that the idea of a universal basic income could find a serious hearing in the United States if we followed the example of other countries considering the experiment, like Finland and Canada. (In November 2015, the Finnish government proposed paying 800 euros per month to every adult.) Among other benefits, a universal basic income or UBI would be a way to compensate for all that unpaid care work, so often done by women.[42]

At a moment when the "unfinished business" of reforming the workplace is getting some much needed attention—and some innovative new approaches—the high tech option of egg freezing seems ironically regressive.

Justice is ultimately about giving everyone their due, not only the privileged few with access to egg freezing. Ultimately, I think affirmative action is not the right analogy for egg freezing. I think a better analogy would be to say egg freezing is like giving—or *selling*—iPads to a few gifted third graders when the entire school lacks the structure to effectively teach all of its students to read. Surely, some women will benefit if they freeze their eggs when they are young enough and they beat the odds when it's time to thaw and fertilize. But egg freezing's role in the feminist quest for equality in the workplace is marginal, at best.

Endnotes

1. The Practice Committees of the American Society for Reproductive Medicine and the Society for Assisted Reproductive Technology, "Mature Oocyte Cryopreservation: A Guideline," *Fertility and Sterility* 99, no. 1 (January 2013): 37–43.
2. ASRM and SART, "Mature Oocyte Cryopreservation," 42.
3. An "EggBanxx cycle plan" includes cycle monitoring, cycle management, cryopreservation of eggs with vitrification, one year of cryopreserved storage, and egg retrieval. EggBanxx, accessed April 5, 2016, https://www.eggbanxx.com/
4. Jennifer Ludden, "Nudging Young Women to Think About Fertility," National Public Radio. May 31, 2011. Accessed April 5, 2016, http://www.npr.org/2011/05/31/136401095/nudging-young-women-to-think-about-fertility

5. Extend Fertility, accessed April 5, 2016, http://www.extendfertility.com/how-it-works/pricing
6. Abby Rabinowitz, "Why Egg Freezing is an Impossible Choice," *Nautilus*, March 19, 2015. Accessed April 5, 2016, http://nautil.us/issue/22/slow/why-egg-freezing-is-an-impossible-choice
7. Centers for Disease Control and Prevention. "2013 Assisted Reproductive Technology Fertility Clinic Success Rates Report," October 2015. Accessed April 5, 2016, ftp://ftp.cdc.gov/pub/Publications/art/ART-2013-Clinic-Report-Full.pdf#page=9
8. Personal correspondence received from the CDC on March 16, 2016: "As you are probably aware, egg freezing has only recently been designated as non-experimental (October 2012). As such, we started collecting this information with the 2013 data. These cycles don't fit into any of our traditional categories for reporting success rates (fresh/frozen embryo, donor/nondonor) so we are still trying to figure out the best way to present this information. Please note that we currently report the total number of cycles using frozen eggs in Fig. 2 of the National Summary Report (2013). We also hope to add a figure depicting national level success rates from frozen egg cycles in the future."
9. CDC, "2013 ART Report."
10. Rabinowitz, "Impossible Choice." According to Rabinowitz's interview with Noyes: "Using frozen eggs, the NYU Fertility Center had achieved 62 births and ongoing pregnancies—a 44% success rate by cycle…."
11. The American College of Obstetricians and Gynecologists Committee on Gynecologic Practice. "Committee Opinion: Oocyte Cryopreservation," *Obstetrics and Gynecology* 123, no. 1 (2014): 221–222.
12. ASRM and SART, "Mature Ooctye Cryopreservation," 41.
13. Heidi Mertes and Guido Pennings, "Social Egg Freezing: For Better Not for Worse," *Reproductive Biomedicine Online* 23, no. 7 (2011): 824–829.
14. T.J. Mathews and Brady E. Hamilton, "Mean Age of Mothers is on the Rise: United States, 2000–2014," National Center for Health Statistics Data Brief, No. 232, January 2016. Accessed April 5, 2016, http://www.cdc.gov/nchs/data/databriefs/db232.htm
15. T.J. Matthews and Brady E. Hamilton, "Delayed Childbearing: More Women Are Having Their First Child Later in Life," National Center for Health Statistics Data Brief, No. 21, August 2009. Accessed April 5, 2016, http://www.cdc.gov/nchs/data/databriefs/db21.pdf
16. Matthews and Hamilton, "Delayed Childbearing."
17. Anne-Marie Slaughter, *Unfinished Business: Women, Men, Work, and Family* (New York: Random House, 2015), 232. According to Slaughter, nearly 60% of women are in the workforce. See also United States Department of Labor, Data and Statistics, "Women in the Labor Force," which states 70% of women with children under 18 participate in the labor force. Accessed April 5, 2016, http://www.dol.gov/wb/stats/stats_data.htm
18. Ann Crittenden, *The Price of Motherhood: Why the Most Important Job in the World is Still the Least Valued*. New York: Metropolitan Books, 2001.
19. Suzanne M. Bianchi, Melissa A. Milkie, Liana C. Sayer, and John P. Robinson, "Is Anyone Doing the Housework? Trends in the Gender Division of Household Labor," *Social Forces* 79, no. 1 (September 2000): 191–228.
20. Melissa Davey, "Women Start Out as Ambitious as Men but it Erodes Over Time, Says Researcher," *The Guardian*, November 19, 2015. Accessed April 5, 2015, http://www.theguardian.com/australia-news/2015/nov/19/women-start-out-as-ambitious-as-men-but-it-erodes-over-time-says-researcher. (The researcher is Michelle Ryan, a professor of social and organizational psychology at University of Exeter in UK.)
21. Robert Reich, *Inequality For All*. Beverly Hills, CA: Anchor Bay Entertainment, 2014.
22. Imogen Goold and Julian Savulescu, "In Favour of Freezing Eggs for Non-medical Reasons," *Bioethics* 23, no. 1 (2009): 47–58.
23. Michael Schmidt, "Pentagon to Offer Plan to Store Eggs and Sperm to Retain Young Troops," *The New York Times*. February 3, 2016. Accessed April 5, 2016, http://www.nytimes.

com/2016/02/04/us/politics/pentagon-to-offer-plan-to-store-eggs-and-sperm-to-retain-young-troops.html?_r=1
24. Goold and Savulescu, "In Favour of Freezing Eggs," 50.
25. Ibid.
26. Ibid.
27. Ibid.
28. Ibid.
29. Emma Rosenblum, "Later, Baby: Will Freezing Your Eggs Free Your Career?" *Bloomberg Business*, April 17, 2014. Accessed April 5, 2016, http://www.bloomberg.com/bw/articles/2014-04-17/new-egg-freezing-technology-eases-womens-career-family-angst
30. Susan Bordo, "Material Girl: The Effacements of Postmodern Culture," *Michigan Quarterly Review* 29, no. 4 (1990): 653–677.
31. Sarah Elizabeth Richards, "Do You Have to Be Rich to Freeze Your Eggs?" *Slate*, August 22, 2013. Accessed April 5, 2016, http://www.slate.com/articles/double_x/doublex/2013/08/the_cost_of_egg_freezing_after_years_of_prohibitive_pricing_clinics_are.html
32. Alison Cooke, Tracey A. Mills, and Tina Lavender, "Advanced Maternal Age: Delayed Childbearing is Rarely a Conscious Choice," *International Journal of Nursing Studies* 49, no. 1 (January 2012): 36.
33. Angel Petropanagos, "Reproductive 'Choice' and Egg Freezing," in *Oncofertility: Ethical, Legal, Social and Medical Perspectives*, ed. Teresa K. Woodruff, Laurie Zoloth, Lisa Campo-Engelstein, and Sarah Rodriguez (New York: Springer, 2010), 230.
34. Petropanagos, "Reproductive 'Choice' and Egg Freezing," 234.
35. Michael Sandel, "The Case Against Perfection," *The Atlantic Monthly*. 293, no. 3 (April 2004): 50–54, 56–60, 62.
36. Seema Mohapatra, "Using Egg Freezing to Extend the Biological Clock: Fertility Insurance or False Hope?" *Harvard Law and Policy Review* 8, no. 2 (2014): 381–411.
37. G Witkin, A Tran, J.A. Lee, L. Schuman, L. Grunfeld, and J.M. Knopman, "What Makes a Woman Freeze: The Impetus Behind Patients' Desires to Undergo Elective Oocyte Cryopreservation," *Fertility and Steril*ity 100, no. 3 (September 2013): S24.
38. Rabinowitz, "Impossible Choice."
39. See, for example, Rebecca Mead, "Cold Comfort: Tech Jobs and Egg Freezing," *The New Yorker*; October 17, 2014. Accessed April 5, 2016, http://www.newyorker.com/news/daily-comment/facebook-apple-egg-freezing-benefits. See also Alana Cattapan, Kathleen Hammond, Jennie Haw, and Lesley A. Tarasoff, "Breaking The Ice: Young Feminist Scholars of Reproductive Politics Reflect on Egg Freezing," *International Journal of Feminist Approaches to Bioethics* 7, no. 2 (2014): 236–247. See also Karey Harwood, "Egg Freezing: A Breakthrough for Reproductive Autonomy?" *Bioethics* 23, no. 1 (2009): 39–46.
40. Susan Dominus, "Rethinking the Work-Life Equation," *The New York Times Magazine*, February 28, 2016, 47–49. Accessed April 5, 2016, http://www.nytimes.com/2016/02/28/magazine/rethinking-the-work-life-equation.html
41. PR Newswire, MIT Sloan School of Management, "First Ever Use of Control Group to Measure Effectiveness of Workplace Flexibility Program Shows Increased Job Satisfaction, Reduced Burnout and Stress say U Minnesota and MIT Sloan Co-Authors," January 13, 2016. Accessed April 5, 2016, http://www.prnewswire.com/news-releases/first-ever-use-of-control-group-to-measure-effectiveness-of-workplace-flexibility-program-shows-increased-job-satisfaction-reduced-burnout-and-stress-say-u-minnesota-and-mit-sloan-co-authors-300203452.html. See also Erin L. Kelly, Phyllis Moen, and Eric Tranby, "Changing Workplaces to Reduce Work-Family Conflict: Schedule Control in a White-Collar Organization," *American Sociological Review* 76, no. 2 (2011): 265–290.
42. Judith Shulevitz, "It's Payback Time for Women," *The New York Times*, January 8, 2016. Accessed April 5, 2016, http://www.nytimes.com/2016/01/10/opinion/sunday/payback-time-for-women.html?smid=pl-share&_r=0

Bibliography

Centers for Disease Control and Prevention (Oct 2015) 2013 Assisted reproductive technology fertility clinic success rates report. ftp://ftp.cdc.gov/pub/Publications/art/ART-2013-Clinic-Report-Full.pdf#page=9. Accessed 5 April 2016

Bianchi SM, Milkie MA, Sayer LC, Robinson JP (2000) Is anyone doing the housework? Trends in the gender division of household labor. Soc Forces 79(1):191–228

Bordo S (1990) Material girl: the effacements of postmodern culture. Mich Q Rev 29(4):653–677

Cattapan A, Hammond K, Haw J, Tarasoff LA (2014) Breaking the ice: young feminist scholars of reproductive politics reflect on egg freezing. Int J Fem Approaches Bioeth 7(2):236–247

Cooke A, Mills TA, Lavender T (2012) Advanced maternal age: delayed childbearing is rarely a conscious choice. Int J Nurs Stud 49(1):30–39

Crittenden A (2001) The price of motherhood: why the most important job in the world is still the least valued. Metropolitan Books, New York

Davey M(19 Nov 2015) Women start out as ambitious as men but it erodes over time, says researcher. The Guardian. http://www.theguardian.com/australia-news/2015/nov/19/women-start-out-as-ambitious-as-men-but-it-erodes-over-time-says-researcher. Accessed 5 April 2015

Dominus S (28 Feb 2016) Rethinking the work-life equation. The New York Times Magazine: 47–49. http://www.nytimes.com/2016/02/28/magazine/rethinking-the-work-life-equation.html. Accessed 5 April 2016

EggBanxx. https://www.eggbanxx.com/. Accessed 5 April 2016

Extend Fertility. http://www.extendfertility.com/how-it-works/pricing. Accessed 5 April 2016

Goold I, Savulescu J (2009) In favour of freezing eggs for non-medical reasons. Bioethics 23(1):47–58

Harwood K (2009) Egg freezing: a breakthrough for reproductive autonomy? Bioethics 23(1):39–46

Kelly EL, Moen P, Tranby E (2011) Changing workplaces to reduce work-family conflict: schedule control in a white-collar organization. Am Sociol Rev 76(2):265–290

Ludden J31 May 2011 Nudging young women to think about fertility. National Public Radio. http://www.npr.org/2011/05/31/136401095/nudging-young-women-to-think-about-fertility

Matthews TJ, Hamilton BE (Aug 2009) Delayed childbearing: more women are having their first child later in life. National Center for Health Statistics Data Brief, vol 21. http://www.cdc.gov/nchs/data/databriefs/db21.pdf. Accessed April 5, 2016

Mathews TJ, Hamilton BE (Jan 2016) Mean age of mothers is on the rise: United States, 2000–2014. National Center for Health Statistics data brief, vol 232. http://www.cdc.gov/nchs/data/databriefs/db232.htm. Accessed 5 April 2016

Mead R (17 Oct 2014) Cold comfort: tech jobs and egg freezing. The New Yorker. http://www.newyorker.com/news/daily-comment/facebook-apple-egg-freezing-benefits. Accessed 5 April 2016

Mertes H, Pennings G (2011) Social egg freezing: for better not for worse. Reprod Biomed Online 23(7):824–829

Mohapatra S (2014) Using egg freezing to extend the biological clock: fertility insurance or false hope? Harv Law Policy Rev 8(2):381–411

Petropanagos A (2010) Reproductive 'choice' and egg freezing. In: Woodruff TK, Zoloth L, Campo-Engelstein L, Rodriguez S (eds) Oncofertility: ethical, legal, social and medical perspectives. Springer, New York, pp 223–235

PR Newswire, MIT Sloan School of Management (13 Jan 2016) First ever use of control group to measure effectiveness of workplace flexibility program shows increased job satisfaction, reduced burnout and stress say U Minnesota and MIT Sloan Co-Authors. http://www.prnewswire.com/news-releases/first-ever-use-of-control-group-to-measure-effectiveness-of-workplace-flexibility-program-shows-increased-job-satisfaction-reduced-burnout-and-stress-say-u-minnesota-and-mit-sloan-co-authors-300203452.html. Accessed 5 April 2016

Rabinowitz A (19 Mar 2015) Why egg freezing is an impossible choice. Nautilus. http://nautil.us/issue/22/slow/why-egg-freezing-is-an-impossible-choice. Accessed 5 April 2016

Reich R (2014) Inequality for all. Anchor Bay Entertainment, Beverly Hills
Rosenblum E (17 April 2014) Later, baby: will freezing your eggs free your career? Bloomberg Business. http://www.bloomberg.com/bw/articles/2014-04-17/new-egg-freezing-technology-eases-womens-career-family-angst. Accessed 5 April 2016
Sandel M (April 2004) The case against perfection. The Atlantic Monthly 293(3):51–62
Shulevitz J (8 Jan 2016) It's payback time for women. The New York Times. http://www.nytimes.com/2016/01/10/opinion/sunday/payback-time-for-women.html?smid=pl-share&_r=0. Accessed 5 April 2016
Slaughter A-M (2015) Unfinished business: women, men, work, and family. Random House, New York
Schmidt M (3 Feb 2016) Pentagon to offer plan to store eggs and sperm to retain young troops. The New York Times. http://www.nytimes.com/2016/02/04/us/politics/pentagon-to-offer-plan-to--store-eggs-and-sperm-to-retain-young-troops.html?_r=1. Accessed 5 April 2016
The American College of Obstetricians and Gynecologists Committee on Gynecologic Practice (2014) Committee opinion: oocyte cryopreservation. Obstet Gynecol 123(1):221–222
The Practice Committees of the American Society for Reproductive Medicine and the Society for Assisted Reproductive Technology (2013) Mature oocyte cryopreservation: a guideline. Fertil Steril 99(1):37–43
U.S. Bureau of Labor Statistics. Women in the labor force: a databook (December 2015) BLS Reports 1059:1–105. https://www.bls.gov/opub/reports/womens-databook/archive/women-in-the-labor-force-a-databook-2015.pdf. Accessed March 10, 2017
Witkin G, Tran A, Lee JA, Schuman L, Grunfeld L, Knopman JM (2013) What makes a woman freeze: the impetus behind patients' desires to undergo elective oocyte cryopreservation. Fertil Steril 100(3):S24

Fertility Preservation in Transgender and Gender-Nonconforming Youth and Adolescents

Steph deNormand

As medical transition for transgender people has become more widely available, treatments and the ethics surrounding them have become an area of significant debate. Social awareness of trans[1] people and their specific needs have fueled a movement towards more inclusive healthcare and access to potentially lifesaving gender-affirming treatments, such as hormone therapies and surgeries. Many of these treatments may result in infertility, potentially creating additional challenges having children later in life. Since fertility preservation options are generally considered elective procedures for fertile people of reproductive age, they are often only available to those who can afford this care. Additionally, these technologies are rarely made available, or even discussed, when the individual transitioning has not yet reached puberty. In order to outline the needs of these individuals, this chapter uses the example of fertility preservation for young oncology patients as an analogous situation to that of transgender and gender-nonconforming youth who are facing a decrease or potential loss of fertility. Here, in combination with discussions of bioethics and biomedicine with feminist and trans theory, an analysis of this particular population's rights to reproduction (and therefore fertility preservation) has a clear and distinct place among bioethical literature and constitutes an area in which continued investigation is necessary.

Trans identities, while recently benefitting from increased media attention, are still widely misunderstood by the general population and a large portion of the medical field. In order to ground this analysis there are a few essential assumptions, supported by current literature, which must be made to move forward. The first of

[1] For this discussion, "trans" indicates all individuals who identify as transgender, transsexual, and/or gender-nonconforming. This means that they do not identify with the sex they were assigned at birth, and therefore may aim to "transition" socially or medically to better align their physical and/or social selves with their internal sense of their gender.

S. deNormand (✉)
Women's, Gender, and Sexuality Studies Department, University at Albany, Albany, NY, USA
e-mail: srdenormand@gmail.com; sdenormand@albany.edu

L. Campo-Engelstein, P. Burcher (eds.), *Reproductive Ethics*,
DOI 10.1007/978-3-319-52630-0_6

these assumptions is that trans people exist separate of a medicalized identity, and that the proper treatment of these individuals constitutes easing gender dysphoria. This can be accomplished through a variety of interventions, and whether social transition, psychotherapy, hormone therapy, surgery, or any combination of these treatments are included in this care should be determined on an individual basis (Coleman et al. 2012: 171). While therapeutic options have historically included psychotherapies with the intention of aligning the individual's gender[2] with their sex assigned at birth, these treatments are not considered ethical by current standards. Psychotherapies should instead be focused on "reducing … distress related to the gender dysphoria and on ameliorating any other psychosocial difficulties" (Coleman et al. 2012: 175). Many of these therapies however, including hormonal and surgical, result in or have the potential to result in infertility. This may represent a significant loss for the individual, but is generally seen as a necessary side effect of these care options (T'sjoen et al. 2013: 575).

It is also important to assert that there is desire for this analysis by the affected population; that transgender and gender-nonconforming people have interest in having children using their own genetic material. This desire has been made apparent through multiple studies and constitutes transphobic assumptions of these individuals to assume otherwise (Nixon 2013: 94). Research conducted within the past 5 years indicates that "many transgender persons are of reproductive age at the time of transition, and confirms that many may wish to have children after transition" (Ethics Committee 2015: 1112). Individuals who identify openly as transgender or gender-nonconforming, or who appear to deviate from a heteronormative family structure, have been historically discriminated against in access to assisted reproductive technologies (ARTs), as well as other ways of creating a family such as adoption (Ethics Committee 2015: 1112). Additionally, there is no indication that transgender people are less suited to be parents. The American Academy of Child and Adolescent Psychiatry released an assessment stating "there is no evidence to support that parents who are … transgender are per se deficient in parenting skills, child-centered concerns, and parent-child attachments compared with heterosexual [and presumably cisgender] parents" (Ethics Committee 2015: 1112).

It is important to note the interlocking oppressions present within populations which lack privilege. Racism and classism are just two of the potential social disadvantages that a trans person may experience in addition to cissexism[3]; however, classism is particularly of note in this analysis. Medical transition itself is expensive, and currently even those who can afford insurance are not guaranteed to have any particular aspect of their treatment covered. Fertility preservation represents an additional cost, which varies by treatment option but can cost hundreds to thousands

[2]Gender represents the internal sense of self one has in reference to the socially constructed roles of men and women. When an individual's gender is in line with their sex assigned at birth they are cisgender, and when their gender is not in line with their sex assigned at birth, they may identify as transgender.

[3]Cissexism is the belief that transgender people or their unique needs are inherently inferior to or less important than those of cisgender people.

of dollars for only potentially viable cells, and additional costs for the use of those cells (Snyder and Tate 2013: 175). This cost represents another example of transphobia, and is based on the assumption that hormonal and surgical treatments for gender dysphoria are elective, and therefore offering fertility preservation is seen as optional. As a result, only individuals who can afford to take on these additional costs will have potential access to seek out these resources.

In line with some of the arguments made around oncofertility,[4] individuals undergoing hormonal or surgical transition should not only be informed of and offered fertility preservation options, but they should also have these options covered by health insurance companies. In 2013, "the American Medical Association adopted a measure to support legislation that would require health insurers to cover fertility preservation when cancer treatments could result in infertility," on the grounds that fertility preservation represents "an essential part of the management of their cancer" (Nixon 2013: 96). For people seeking out hormonal and surgical treatments for gender dysphoria in which fertility is affected, discussions around fertility are similarly essential, and therefore should also be supported and covered by health insurance companies. "Young transgender people should not have to forego the prospect of future children in order to obtain certain hormone therapies and gender-confirming surgeries to alleviate their gender dysphoria" (Nixon 2013: 102). While having these options covered by insurance does not remove all classist implications, it does begin the conversation around access to fertility treatments, and would increase access considerably from the current out-of-pocket standard.

There are multiple issues that are commonly brought up in discussions around fertility treatment and preservation in transgender youth. Through the use of a reproductive justice framework, as well as bioethical and biomedical studies surrounding issues of fertility preservation and transgender care, these concerns can be understood and reimagined in order to make available these important technologies and allow transgender people to have biological children.

The first topic frequently brought into these discussions is that children and adolescents are either unable or unwilling to make decisions about their future reproduction, and that these young people are below the age of consent for an elective procedure such as tissue donation for preservation. This concern has multiple layers of complexity, and therefore must be broken down accordingly. This controversy can be entirely avoided by allowing the individual to reach the age of consent before making decisions surrounding fertility. Currently, medications such as GnRH analogues, commonly referred to as "puberty blockers" or simply "blockers," are used in order to suppress endogenous puberty in transgender and gender-nonconforming youth (Khatchadourian et al. 2014: 908). In this context, endogenous puberty is considered to be when an individual undergoes puberty according to the sex they were assigned at birth, which would typically result in a phenotypic presentation in line with their sex assigned at birth. Blockers allow for endogenous puberty to be halted, and are typically administered at or after Tanner Stage II (Coleman et al.

[4] Oncofertility is the use of fertility preservation and reproductive technologies in individuals undergoing cancer treatments.

2012: 177). This treatment is considered reversible since the individual will continue to undergo endogenous puberty if they stop participating in this treatment, and both treatment and cessation of blockers have relatively minimal side effects. Since feminizing and masculinizing hormone therapies are typically not started until the individual reaches the age of consent either, usually around age sixteen, this appears to completely avoid concerns of consent to treatment surrounding reproduction.

Occasionally, and particularly surrounding reproduction, the age of consent is still considered too young to be making the potentially life-changing decision such as whether or not they want biological children, and therefore to undergo this procedure. This is a fallacious argument however, particularly if the individual is seeking out feminizing or masculinizing hormone therapy after going through with fertility preservation. By making the decision to begin this form of hormone treatment, transgender and gender-nonconforming individuals make the decision to become temporarily infertile (while actively undergoing the therapy) with the potential of resulting in permanent infertility. By taking this as a known risk, transitioning individuals are making the choice to become infertile. Therefore, if this person would prefer to undergo fertility preservation, they are in fact providing themselves more options for their future, not fewer.

Potential for convincing or coercion from parents or guardians has been another significant area of concern. This could be a very real problem, particularly for trans men and gender-nonconforming individuals who were assigned female at birth. Often people who are perceived to have female bodies are expected to have a strong desire for biological children. This may present itself as an additional pressure on trans men, and particularly for those who do not want to seek out fertility preservation. Parents, guardians, or even physicians may feel the desire to encourage egg donation, whether it be out of concern for the trans person or out of selfish desires, such as the desire to be a grandparent. The simple potential for this option to be abused should not constitute a reason for it to not be offered. In fact, this is simply another reason to ensure that the individual is given options which are clearly explained, and that the individual's informed decision is honored. Recommendations from care providers such as mental health professionals, physicians, and endocrine or fertility experts should be obtained as necessary; however the final decision to participate in this process should be left to the individual. Additionally, conversations and counseling should be considered for the parents and/or guardians of the individual, in order to make sure that they are also well informed and equipped to support them.

The above concerns can be additionally countered by stating that these forms of fertility preservation are already being performed in cases of gonadotoxicity, such as in cancer treatments. The field of oncofertility has become a fertile area for this research, and has included studies involving people at multiple stages of life. Fertility preservation has been discussed as an essential part of cancer treatment, and it has become standard to offer fertility preservation options in preparation for future infertility or sterility in these treatments. These preservation options are offered to individuals as young as 2 years old, in which case they have far greater potential for parent's interests being considered over that of the child and are far less likely to have the child's informed decision taken into account (Quinn et al. 2012: 38). Therefore denying any young individual the ability to use fertility preservation technologies, including not informing them of these options as a denial of the ability to choose, constitutes discrimination based on the individual's gender identity and decision to seek out medical transition.

The only other difference between the cases dealt with in oncofertility studies and in transgender youth is the issue of gender dysphoria. In order to undergo classical egg or sperm collection, an individual must reach a particular level of reproductive maturity. In individuals who undergo fertility preservation and medical transition after completing endogenous puberty, this is typically not a problem barring any additional fertility difficulties. If an individual chooses to take puberty blockers, these organs never fully develop, and as a result this classical form of collection is likely not an option (Coleman et al. 2012: 177). The argument is therefore often made that the mental toll of experiencing endogenous puberty is too great for someone who is transgender or gender-nonconforming, and should not be considered an option. While this is a legitimate concern, there are forms of collection that do not follow the classical ways, another area which has been thoroughly researched through pediatric oncofertility. These are still experimental procedures, and therefore some caution on the behalf of the physician is understandable if not warranted. However, there is still an essentialized assumption in this argument about experiences of dysphoria; particularly that it is experienced similarly for everyone who is transgender or gender-nonconforming. This is simply not the case, and a discussion about the potential benefits of undergoing a portion of endogenous puberty should be had in order to be sure that the individual is fully informed. If their gender dysphoria presents in a way that is manageable for them short term, or if they value stronger reproductive options over this struggle, they should be allowed to make that decision. This is not to say that counseling and mental health therapies should not be involved, and in fact these choices should be made with the support of a mental health professional; but the final, fully informed decision should be for to the transgender or gender-nonconforming individual to make for themselves.

Based on this analysis, a variety of recommendations can be made in relation to fertility preservation for youth and adolescents seeking out physical transition through hormone therapies and/or surgeries. There is importantly no treatment standard aside from providing information and options to the individual, because they must have the right to make decisions in regards to their own body and in accordance with their own personal experience with gender dysphoria. As such, fertility preservation should never be done without the person's consent, and should always be considered a process one must "opt in" to, as opposed to "opt out" of. Through an "opt in" treatment plan as conceived here, the transitioning individual would be able to stop any treatment at any time, for any reason (while following medical safety standards). Included in this would be any unanticipated effects of treatment, or changes in desire to preserve fertility. This allows for a plan that the individual can feel secure in, and would allow them to feel validated and supported in any decisions they make throughout their care. Below are two potential routes this care could take, which each separately address the concerns that have been raised about fertility treatment for transitioning youth. It is also important to note that these interventions are based on the assumption that the transitioning youth has articulated their desire to transition prior to reaching Tanner Stage II of sexual development. If this is not the case, the routes below would need to be modified based on the potential remaining effectiveness of puberty blockers, the degree of endogenous puberty that has already taken place, and the age of the individual in relation to the age of consent.

Route 1: Egg or Sperm Collection upon Reproductive Maturity.

This allows the individual to develop to reproductive maturity without the use of hormone or surgical therapies, which would allow for classical sperm or egg donation methods at this time, and then beginning interventions such as hormone therapy or surgery after this time. The largest disadvantage to this problem is that it requires a certain amount of puberty to occur as a result of endogenous hormones, and may result in significantly increased dysphoria. While this is a distinct disadvantage, it is important that this is an option that is articulated to the transitioning youth. Gender dysphoria is experienced differently by different individuals, and the potential benefit of having viable reproductive tissue may outweigh potential dysphoria experienced. It is also important in this method that the treatment provider explain the level of sexual development that the person must reach in order to donate, particularly describing what gendered aspects of puberty (such as deepening of voice, hair growth, or breast tissue growth) will or will not be undone through feminizing or masculinizing hormone therapies (Coleman et al. 2012: 188).

Route 2: Use of Puberty Blockers until "Adulthood."

This option specifically addresses concerns surrounding youth's ability to make decisions about their future, particularly surrounding potential sexuality or parenthood. Puberty blockers (such as GnRH analogues) can begin to be administered following the Standards of Care, which recommends waiting until the individual reaches Tanner Stage II (Coleman et al. 2012: 177). Since this can begin as young as age nine, puberty blockers have been used to allow for time to explore and develop their gender identity. This time could also be used to develop an opinion on fertility treatments, as feminizing and masculinizing hormone therapies are typically not offered until the transitioning individual reaches the age of sixteen, or the local age of consent (Coleman et al. 2012: 178). At this point, the individual could make the decision to undergo endogenous puberty and egg or sperm donation as described in Route 1, or to seek out feminizing or masculinizing hormone therapy and/or surgery to continue their physical transition.

There are a few additional considerations and potential options that may be more appealing to transitioning individuals, however are considered potentially less effective or are still experimental procedures. One such procedure is the collection and preservation of either immature testicular or ovarian tissue. This would be a potential option for any transitioning individual prior to beginning treatment with feminizing or masculinizing hormone therapies, and would not require endogenous puberty to take place. This technology is still experimental; however it has been an area of continued research in oncofertility and where fertility may be affected by gonadotoxic therapies. While there are currently no human cases of this fertility preservation resulting in successful human embryos, animal testing has been promising, and both immature testicular and ovarian tissue has been collected from pediatric oncology patients, and other pediatric patients undergoing gonadotoxic therapies (Wyns et al. 2010: 312; Quinn et al. 2012: 38). Given this precedent, there is no reason that these methods could not also be an option for transgender youth, as this procedure is being done to retrieve ovarian tissue (the more invasive of the two operations) as young as 2 years of age (Quinn et al. 2012: 38). While this is not guaranteed to be a viable option, it may represent an acceptable middle ground for youth who do not want to undergo endogenous puberty, but would still prefer to have some potential for children from their genetic material.

Another area of potential interest is the ability to collect sperm and eggs after having been on feminizing or masculinizing hormones. This research has shown to work in both transgender women and men. However, it has only been conducted on people with mature testicular or ovarian tissue (Gidoni et al. 2013: S170; Coleman et al. 2012: 197). This option is more suited to individuals who have been using feminizing or masculinizing hormone therapies for a shorter period of time, decreasing the likelihood of lasting effects impacting the specific tissue (Rodriguez-Wallberg et al. 2014: e160). Depending on the individual's particular gender dysphoria and desire for reproductive options, this may be an ideal option for some, particularly those who have only been on feminizing or masculinizing hormones for a few years. Further research should be done in order to determine if this is a viable option for those who have not fully undergone endogenous puberty.

A final area of consideration and current research is the possibility of uterine transplants for transgender women and transfeminine people. While this is not strictly an issue of fertility preservation, the potential for pregnancy can be an essential part of an individual's conception of femininity and motherhood, and therefore may be an important reproductive option. There have been limited studies on uterine transplants in cisgender women, which have been successful for a sufficient amount of time for a pregnancy (Ozkan et al. 2013: 473). This area of technological and surgical advancement has profound implications for some transgender women, and will likely add additional considerations to their reproductive choices once this procedure has been more thoroughly researched for this population, and on the effects that may result from transplanting into someone who was male assigned at birth.

All issues considered, transgender youth should have just as much of a right to reproductive justice and freedom as any other individuals. While there are certainly ongoing concerns about access to treatments and therapies for transgender individuals experiencing multiple layers of oppression, continued conversation is also needed with and in relation to younger populations. As technological advances create more possibilities, it is important that we continue to assess the unequal ways in which these technologies are made available, and the biases that continue to be expressed against transgender and gender-nonconforming individuals.

References

Coleman E, Bockting W, Botzer M, Cohen-Kettenis P, Decuypere G, Feldman J, Fraser L, Green J, Knudson G, Meyer WJ, Monstrey S, Adler RK, Brown GR, Devor AH, Ehrbar R, Ettner R, Eyler E, Garofalo R, Karasic DH, Lev AI, Mayer G, Meyer-Bahlburg H, Hall BP, Pfaefflin F, Rachlin K, Robinson B, Schechter LS, Tangpricha V, Van Trotsenburg M, Vitale A, Winter S, Whittle S, Wylie KR, Zucker K (2012) Standards of care for the health of transsexual, transgender, and gender-nonconforming people, Version 7. Int J Transgenderism 13(4):165–232

Ethics Committee of the American Society for Reproductive Medicine (2015) Access to fertility services by transgender persons: an Ethics Committee opinion. Fertil Steril 104(5):1111–1115
Gidoni YS, Raziel A, Strassburger DF, Kasterstein E, Ben-Ami I, Ron-El R (2013) Can we preserve fertility in a female to male transgender after a long term testosterone treatment—case report. Fertil Steril 100(3):S169–S170
Khatchadourian K, Amed S, Metzger D (2014) Clinical management of youth with gender dysphoria in vancouver. J Pediatr 164(4):906–911
Nixon L (2013) The right to (trans) parent: a reproductive justice approach to reproductive rights, fertility, and family-building issues facing transgender people. William Mary J Women Law 20(1):73–103
Ozkan O, Akar ME, Erdogan O, Hadimioglu N, Yilmaz M, Gunseren F, Cincik M, Pestereli E, Kocak H, Mutlu D, Dinckan A, Gecici O, Bektas G, Suleymanlar G (2013) Preliminary results of the first human uterus transplantation from a multiorgan donor. Fertil Steril 99(2):470–476
Quinn GP, Stearsman DK, Campo-Engelstein L, Murphy D (2012) Preserving the right to future children: an ethical case analysis. Am J Bioethics 12(6):38–43
Rodriguez-Wallberg KA, Dhejne C, Stefenson M, Degerblad M, Olofsson JI (2014) Preserving eggs for men's fertility. A pilot experience with fertility preservation for female-to-male transsexuals in Sweden. ASRM Abstr 102(3):E160–E161
Snyder KA, Tate AL (2013) What to do now? How women with breast cancer make fertility preservation decisions. J Fam Plann Reprod Health Care 39(3):172–178
T'sjoen G, Caenegem EV, Wierckx K (2013) Transgenderism and reproduction. Curr Opin Endocrinol Diabetes Obes 20(6):575–579
Wyns C, Curaba M, Vanabelle B, Van Langendonckt A, Donnez J (2010) Options for fertility preservation in prepubertal boys. Hum Reprod Update 16(3):312–328

Part III
Childbirth

Rights and Realities in U.S. Maternity Care

Hermine Hayes-Klein

Introduction

The healthcare that women receive during pregnancy and childbirth implicates reproductive choice, bioethics, human rights, and feminism. The elimination of preventable maternal and neonatal mortality and morbidity requires not only that women can access skilled healthcare and emergency support for pregnancy, childbirth, and postpartum, but that the care women receive is culturally acceptable to them, and respects their dignity and autonomy. Women can experience labor and childbirth as a time of extreme vulnerability as well as extraordinary power. The way that they are treated during childbirth can amplify their vulnerability or their strength, and leave them feeling traumatized or empowered. In the last few years, grassroots groups in nations around the world have been organizing to speak out about pregnant people's experiences of maternal healthcare, catalyzing the emerging global recognition of the mistreatment in childbirth as a human rights issue.[1]

The recognition of preventable maternal mortality as a matter of human rights as well as public health led to the prioritization of maternal and infant health under the Millenium Development Goals, and investment in improving access to skilled health workers and emergency obstetric services, particularly in the developing world.[2] Yet without explicit recognition of the full range of human rights at stake in pregnancy and childbirth, including but not limited to the right to life and health, developing systems of care have implemented standards of practice that assume that the end goal of a live mother and live baby justifies the means by which that goal is achieved. The abuse and mistreatment of women in the process of facilitating a live birth is rendered invisible and irrelevant. In both developed and developing nations,

H. Hayes-Klein (✉)
Human Rights in Childbirth, Portland, OR, USA

Human Rights in Childbirth, The Hague, The Netherlands
e-mail: hermine@hayeskleinlaw.com

L. Campo-Engelstein, P. Burcher (eds.), *Reproductive Ethics*,
DOI 10.1007/978-3-319-52630-0_7

systems of obstetric care have transformed in a generation, in a massive shift to pharmaceutically induced, augmented, and surgical deliveries. Women's options for supported physiological childbirth,[3] or even vaginal childbirth, have become limited in hospital settings.[4] Reports of forced interventions and abusive encounters with medical providers, and the political barriers to accessing midwifery support and safe out-of-hospital birth, are giving rise to an increasingly vocal feminist movement that calls for women's needs, rights, and voices to be at the center of maternity care policies and practices.

Decision-Making in Childbirth

Many of the dysfunctions that women currently face in maternity care, including disrespect and abuse, could be eliminated through clarification that women's human rights are legal rights that remain in effect during pregnancy and childbirth. The right to autonomy, enshrined as informed consent and refusal, situates women as the central, authoritative decision-maker regarding their care, and yet this right is routinely ignored in maternity care. How much would change in the delivery room if it was understood by everybody present that nothing could be done to a birthing woman without fully and accurately informing her of her clinical condition, her healthcare options, and the risk and benefits of those options, and supporting her in making an uncoerced decision about how she wanted to proceed?[5] It would be transformative, in light of current dynamics in which women are told what is going to happen to them and showed where to sign, or not even told, but just acted upon, women have reported, like an "object" or an "animal."[6]

One in three women in the United States are now having their babies delivered by cesarean section. In the mid-1970s, the cesarean section rate was around 5%. There is no evidence that outcomes have improved as the result of obstetric culture's shift to the new era of surgical delivery, and in fact maternal mortality in the U.S. is rising.[7] Cesarean section without a medical indication triples the mother's risk of dying in childbirth, and creates new forms of risk for her future pregnancies, including very serious conditions like placenta accreta.[8] The lowest cesarean section rates are reported for women who have prenatal care and good nutrition in pregnancy, integrated healthcare that includes both midwives and doctors, and continuous support through labor and birth.[9] The most significant variable that affects a woman's chances of giving birth by cesarean section is the hospital where she gives birth.[10]

A 2013 survey reported that 25% of women who had experienced an induction of labor or a cesarean section felt pressured to accept those interventions.[11] A 2014 study found that women who perceived pressure to have a Cesarean section were more than five times more likely to have a one, more than six times more likely to have one with no medical basis, and nearly seven times more likely to have an unplanned cesarean.[12] 59% of women who received episiotomies did not give consent at all.[13] Finally, 20–38% of women reported that the provider made the "final decision" about whether they would receive a planned cesarean surgery.[14]

Despite the commitment of the American College of Obstetricians and Gynecologists to supporting patient autonomy in childbirth, there exist significant gaps between the ethical principles expressed in their Ethics Committee Opinions and the way that informed consent plays out in reality.[15] This disconnect is fueled, in part, by providers' serious concern with perverse liability mandates that they fear make them legally vulnerable in the event that they support a woman's informed decision, but the birth results in a bad outcome.[16]

Reports on the role of liability pressure in obstetrics express a thematic assumption that providers can protect themselves from liability risk if they impose interventions, including cesarean surgery.[17] A liability rule that inclined doctors toward cesarean delivery might make sense if cesarean surgery carried no risks or costs, and vaginal birth were risky and dangerous. But that is not what the evidence shows. When cesarean surgery is medically needed, it can save lives. But when it is not needed, it carries a long list of risks and costs, including a significantly elevated risk of maternal death.[18] Women in U.S. maternity care are giving birth in environments where providers claim that "liability" compels them to push for a surgical birth that happens to profit and convenience the hospital, but imposes risks on mother[19] and baby,[20] up to and including the risk of death. Obstetric providers currently perceive a "liability" mandate that urges intervention and ignores informed consent and refusal, while failing to incentivize judicious decision-making or health care that optimizes maternal and infant health. Women are subjected to unnecessary surgical deliveries that increase the risks to themselves, their babies, and their future pregnancies, on the assertion that it would be too "risky" for providers or hospitals to support them in vaginal birth.

Obstetric providers recommend intervention on the basis of numerous non-clinical factors, including liability pressure and financial incentives; intervention rates therefore vary widely by provider. Empirical studies show—and doctors confess—that providers and hospitals steer women toward cesareansections not only for clinical reasons, but also for non-medical reasons including financial gain, time convenience, and perceptions of liability pressure.[21] The fact that doctors perform unnecessary surgery for financial gain or time convenience does not prove their collective or individual moral turpitude, but rather their human response to economic incentives. When a provider decides whether to recommend an intervention for a given patient, financial considerations and time-convenience factors likely operate on a subconscious level. While higher costs and longer inpatient stays for surgical deliveries benefit hospitals more directly than individual doctors, these institutional economic forces can translate into imperatives that constrain doctors from providing individualized care, or into a medico-cultural argument that "this is the way we do it around here." On a macro level, these forces play out in significantly higher cesarean section rates in for-profit medical settings around the world.[22]

Doctors' recommendations for intervention, including cesareansection, are colored also by their own perspective and values.[23] Studies show that obstetricians choose cesarean section deliveries for themselves in higher numbers than the general population,[24] and are more likely to undervalue physiological birth while considering cesarean delivery a good solution to "perceived labor and birth problems."[25]

If providers believe that cesarean delivery is a good choice and vaginal birth is unnecessary and undesirable, their counseling of patients may be colored by the belief that the refusal of surgery is an unnecessary choice for the "procedure" of supported vaginal birth.

The multiplicity of factors that influence each obstetric provider's decision-making process are reflected in the significant variability of protocols and intervention rates across states, hospitals, and individual doctors. Studies show cesarean section rates ranging from 7.1 to 69.9% across U.S. hospitals.[26] These variations are not reflected in differences in maternal diagnoses or pregnancy complexity of individual patients.[27] Maternity care's variability of practice and ubiquitous overuse of interventions that benefit the provider at the patient's expense might reasonably lead an informed consumer to actively exercise her right to informed consent and refusal as she navigates the health care system. A birthing woman's legal right to authority in the decisions about her care is her only shield against variability and dysfunction in maternity care. All participants bring a constellation of issues, values, and experiences into their decisions, but informed consent and refusal means that the woman, like all health care patients, has the right to weigh all the factors at stake and make the final call. "Shared decision-making," though an admirable goal in the pursuit of patient-centered care, is a misleading; while the provider can share information, advice, and conversation with their patient, the decision about whether to accept a medical intervention belongs, legally, to the patient alone.

Just as providers and healthcare institutions bring a multiplicity of factors and priorities into shared decision-making, maternity care consumers make their healthcare decisions on the basis of many individual considerations and variables. Healthcare systems can expect that, given the diversity of cultural and religious values, financial circumstances, family support, and past experiences that each pregnant patient will bring to labor and delivery, they may see ten different women with similar clinical charts make ten different sets of choices with regard to pain relief, companionship for labor, induction, augmentation, or planned cesarean delivery. Decision-making in childbirth is more complex than it may be in some other areas of medicine, involving complex balancing of potential short and long-term risks, when it is often difficult to say with certainty whether an intervention is needed. The right to informed consent and refusal requires that systems anticipate the variability that women may bring to their decisions in childbirth, and be prepared to meet and support women's individual needs, instead of requiring women to compromise their right to be supported in their healthcare decisions by conforming to one-size-fits-all standards of care.

Human beings, like all mammals, need to feel safe in order to give birth.[28] The hormonal physiology, and the role of psychology, in labor and childbirth are not taught in medical or sex education, and are often unrecognized by the obstetric paradigm and therefore underserved and undervalued in facility-based birth. Childbirth as a physiological process is powered by hormones, including oxytocin, the hormone associated with love and human connection.[29] Those hormones are strongly affected by the birthing women's emotions, which are in turn affected by her perception of whether she is giving birth in a safe environment.[30] Women who give birth in circumstances that meet their mammalian needs for childbirth—circumstances that meet their needs for privacy and ensure that they feels safe, respected, and supported—have the best chance for an uncomplicated labor and birth.[31] In contrast, if a female mammal feels threatened, inhibited, or unsafe during labor, the release of adrenalin can slow or stop the progress of labor, giving rise to potentially dangerous complications.[32] After an un-medicated birth, mother and baby spike the highest level of oxytocin that the human mammal can produce, resulting in a feeling of bliss, euphoria, and love that ensures their bonding and the baby's survival.[33]

These facts about the psychology and physiology of childbirth cast the assumptions and practices of "normal" birth in the U.S. and most of the world into serious question. Why do our systems of care, even in hospital settings, ignore laboring women's mammalian needs, subjecting them instead to bright lights and the gaze of strangers? There is no reason why maternity care cannot be constructed in a way that meets each woman's needs to feel safe during birth. And yet, in many systems around the world, women must sacrifice feeling safe, and even being safe, in order to access the safety that medical backup provides. Widespread reports of disrespect and abuse, dehumanized, traumatizing treatment, and violations of women's rights to privacy and dignity show that women are giving birth in environments in which they cannot feel safe, because in fact, they are not safe.[34] In both the developed and developing world, some women avoid going to the hospital during childbirth, even if it is their only option for healthcare support, because they anticipate mistreatment or neglect.[35] As the risk of giving birth in the hospital rises, women are willing to accept a higher level of risk in giving birth outside the hospital, and women perceive disrespect and abuse as a salient risk factor.

The hormonal physiology of childbirth also suggests that a reason to pursue a normal physiological labor and birth isn't only to avoid unnecessary interventions and the risks they bear, but to access the physiological benefits of unmedicated childbirth. The postpartum spike of oxytocin, and the bliss described by women who give birth in circumstances that enable that spike,[36] indicated that giving birth without injections and intervention wasn't just needless suffering and martyrdom, but held a significant physiological reward that might even be critical to ensuring the strongest start to motherhood and avoiding postpartum depression.[37]

The history of obstetrics reveals that the protocols that many assume are necessary to safe childbirth, like giving birth on the back in stirrups, never had anything to do with women's needs, but were based on gendered, racist old values, starting with a dynamic that put the woman on her back in bed on the basis of the male provider's convenience and the assumption that the lithotomy position preserved "modesty" and that only a "savage" woman would squat.[38] Obstetric technologies have evolved over time, from the common use of general anesthesia and twilight sleep to render women oblivious for delivery, to the age of the "awake and aware" epidural, planned surgical birth, and electronic fetal monitoring (EFM), but these technologies were often implemented without good evidence that they served mothers and babies or optimized healthy outcomes.

The ubiquitous use of EFM is the perfect example of how medico-legal dysfunctions operate to subject women to medical interventions without an evidence basis.[39]

Despite consistent ongoing studies showing that EFM does not improve outcomes and triples the risk of cesarean section, EFM use has become ubiquitous in hospital settings, and many women are told that its use is "hospital policy" that they cannot refuse.[40] The imposition of EFM on birthing women is reinforced and justified by perverse liability incentives based on junk science and perpetuated by a multi-million dollar legal industry that profits from blaming cerebral palsy on the provider's failure to intervene and "deliver" the baby earlier.[41]

Only if courts hold providers liable for violations of women's right to informed consent and refusal in maternity care and, moreover, impose damage awards that recognize the individual and social significance of the harm, will doctor-patient dynamics in obstetrics be liberated from perverse incentives, and reorient toward woman-centered care. At the same time, with the consumer's right to informed consent comes responsibility for the decisions of care. Providers deserve assurance that their responsibility ends where the patient's rights begin. Legal reinforcement of informed consent and refusal must cut both ways: just as courts must find liability for violations of women's right to consent on the basis of information and advice, courts must also protect doctors from liability in cases where they are blamed for a woman's informed choice. Decisions that hold providers liable for a woman's informed decisions undermine the right to consent for all patients and leave doctors vulnerable for providing respectful support.[42] When doctors share their knowledge of risks and benefits, and support women in the decision at stake, they must not later be found liable for that decision on the theory that the patient lacked the expertise to assess and understand the risk. Informed consent and refusal rests upon the assumption that, despite the esoteric nature of medical knowledge, ordinary people can assess their medical alternatives and make decisions about them—including the decision to go against their doctor's advice.

Reproductive Rights and Equality in Collaborative Maternity Care

Although the right of informed consent and refusal assures birthing people's rights to make the decisions about their medical treatment, and gives them the right to walk out of a hospital at any time, the right to say "no" doesn't ensure that active support that safe birth requires. The right to supportive healthcare for physiological birth, out-of-hospital birth, or even vaginal birth after cesarean, implicates the right to privacy and reproductive ethics.

In 2010, the European Court of Human Rights affirmed that the human right to private life includes the right to choose the circumstances of giving birth. In *Ternovszky v. Hungary*,[43] the human rights tribunal made clear that this right includes the choice between giving birth with a midwife or with a doctor, at home or in the hospital. The Court held that governments violate this human right if they fail to regulate out-of-hospital birth in a way that legitimizes it as a healthcare choice and integrates it into the healthcare system, or if birth professionals cannot support women in their maternity care choices without fear of legal sanction. The birth professional with whom the mother in that case, Anna Ternovszky, fought for the right to give birth is Hungarian obstetrician-turned-midwife Agnes Gereb. Despite the Court's decision, the Hungarian government kept Ms. Gereb on house arrest for 3 years for allegations associated with her support of out-of-hospital birth. The state finally released her under the strict condition that she was forbidden to talk to any pregnant woman.

The evidence is conclusive that achieving global targets for the reduction of maternal and perinatal mortality will require healthcare teams that include doctors, nurses, and midwives.[44] There is a growing call for, and movement toward, partnership and integration of midwifery and medicine in the care of pregnant women.[45] The picture of partnership and integration between medicine and midwifery is, however, a patchwork across Europe, across the United States, and around the world, in both law and practice. The maternity care systems with strong, supported midwifery professions that work with reliable backup from emergency obstetric providers have the lowest intervention rates and the best outcomes.[46] Systems without a supported midwifery profession, or where midwives' ability to provide midwifery care is severely circumscribed, increasingly resemble cesarean section assembly lines.[47] The human right described in *Ternovszky* is not yet a reality for most women around the world. On July 21, 2016, the Commissioner for Human Rights for the Council of Europe acknowledged receiving "disturbing reports of human rights violations in the context of maternity health care," and referenced both the *Ternovszky* holding and recent conclusions by the UN Committee on the Elimination of Discrimination Against Women (CEDAW) regarding the "need to ensure… respect for women's rights, dignity and autonomy during deliveries, expressing concerns in particular at reports that childbirth conditions and obstetric services unduly curtail women's reproductive health choices."[48]

Despite the best intentions of maternal health policy makers and providers to move toward the integration of midwifery in maternity care schemes and systems, such integration will be slow in coming without an overriding essential framework of equality and respect between the professions of medicine and midwifery. The fact that the idea of equality between doctors and midwives seems radical is a testament to how entrenched are the systems that have established medicine's dominance over midwifery. The current status of midwives, and the relationships between medicine and midwifery that underlie integration, cannot be understood without recognizing the systemic inequality between obstetric medicine and midwifery, awareness of its historical roots, and a commitment to dismantling that inequality[49] Whether women start birth with a midwife or with a doctor, they need access to emergency obstetric services in the event an emergency arises. Obstetric medicine has used its power to give or withhold emergency services to gain a monopoly over maternity care and to maintain dominance over midwives.

Since time immemorial, women have attended each other in childbirth. Some women developed such skill and expertise that they became the community midwife. Midwives often were the first line of healthcare for the community and tended to the full spectrum of women's reproductive healthcare needs.[50] Midwives held the knowledge of the local herbs and culturally rooted practices associated with fertility, healing, promoting health, and with minimizing the risk of pathology. It was for their knowledge of the herbs for contraception and abortion that midwives were defined as the primary target for the European witch hunts.[51] The torture and execution of midwives over three centuries led to the loss of that knowledge, and a population explosion that fueled many wars and the colonial settlement of the "New World."[52] With the midwives' knowledge of contraception went their knowledge of female sexuality and gynecology, including how to support women in childbirth. After the witch hunts, male doctors stepped in to attend women in childbirth, at first only in emergency cases, but then as the main providers for childbirth, a shift "promoted by the authorities but disliked by women."[53]

> "Up to the late 14th century midwives were entitled to practice without regulation. From this time onward [they] were downgraded from qualified and independent female healers… to mere assistants of the physician [who was] lacking any experience in obstetrics because through the Middle Ages no male was allowed to practice in gynecology."[54]

Despite the knowledge and skills that traditional midwives have always used to serve their communities, there were medical emergencies that could arise in childbirth that they could not solve. Antibiotics, anti-hemorrhagic medicine, assisted and surgical deliveries, and other medical technologies can prevent many of those deaths, and access to such technologies has saved many lives and massively reduced maternal and neonatal loss since their invention. But the terms on which these tools were offered to women, in the U.S. and many other places, created new forms of risk as all women were asked to place themselves in the care of medical professionals for pregnancy and birth, whether or not they needed medical treatment.

In the U.S. and other industrialized nations, state authorities have worked in partnership with medical associations to drive the female population to give birth in hospitals, under the care of doctors.[55] Midwives were often disempowered both as professionals and as citizens, not only on the basis of sex, but also race, immigration status, or caste, in different combinations in different places and times.[56] After black midwives safely delivered generations of babies from both enslaved and free women in the United States, they were degraded through racist propaganda schemes to move women into the hospital, even as women of color in the U.S. had insecure access to hospital care.[57]

Although the subjugation and elimination of midwifery were historically built upon sex inequality, racial inequality, class and caste inequality, and colonial power, like most systems of domination and inequality, it eventually became primarily about capital. Obstetric medicine is a multi-billion dollar industry whose stakeholders include not only providers and hospitals, but pharmaceutical, insurance, and technology industries.[58] After generations of medical monopoly over childbirth, providers and the general public consider it obvious that doctors are superior to midwives, and should have the authority to determine when and how midwives will practice. Many people think that being cared for by a doctor is better than being cared for by a midwife, because that's what they've been always been told. In many places, people don't even know what a midwife is, or think that midwives are from "olden times," and this did not happen by accident.

In the last 40 years, midwifery has re-emerged as a profession, in places where it had disappeared.[59] Regardless of how they are trained and where they work, midwives face powerful structural and systemic barriers to practice and integration. Doctors decide if they are allowed to practice in the hospital and if so, how they practice.[60] The state decides if they are allowed to practice outside the hospital, and often regulates their practice in a way that imposes medical standards on them and prevents them from providing evidence-based and culturally competent midwifery care that upholds their client's human rights.[61] In some places, midwives are still fighting for recognition that midwifery is even a profession.[62] Everywhere, midwifery is underfunded, and in many places, financially unsustainable. Underlying many of the ways that inequality can manifest is that hallmark of discrimination: Disrespect. When medical providers and systems assert dominance over midwives and midwifery, they treat them with disrespect. Not listening, condescending, dismissing, and talking down or rudely to people are all ways of expressing discrimination and disrespect.

What does collaboration based on equality look like, in relations between medicine and midwifery? It looks like mutual respect. In practice, it means that doctors and midwives recognize and respect each other's knowledge and expertise, remain in dialogue to better understand and learn from each other, and have equal voices at the table about maternity care policy. In law, equality requires the recognition of midwifery as an independent and valuable reproductive healthcare profession, and midwives as the authorities in their own standards of care. It also requires subjecting doctors and midwives to fair and equal legal oversight, and acknowledging the role that the law has played historically in marginalizing midwifery and giving medicine a monopoly over maternity care. In finance, equality looks like valuing midwives' role in maternal healthcare, and making sure that midwives and free-standing birth centers get paid as doctors and hospitals do.

The recognition of doctors and midwives as equal and complementary partners in reproductive healthcare would require respect for their relative fields of expertise. Midwives are the experts in physiological birth. Their training teaches them to work with the female body to help women give birth to their babies. As obstetric medicine increasingly relies on surgical delivery, midwives are often the only maternity care providers a woman can find who know how to support vaginal birth, let alone physiological birth, especially in the case of breech or twin deliveries.[63] Doctors' expertise focuses on using medical technologies and treatment to fight pathology and manage medical crises. Medical providers go to school for a long time to learn the complexities of pharmacology and the intricacies of surgery. However, in years of education and residency, an obstetrician might not learn how to support physiological childbirth. Modern obstetric doctors and nurses openly acknowledge that they may never see a woman give birth without intervention in either training or practice, let alone support a physiological breech birth.[64] Improved collaboration between doctors and midwives could operate not only to facilitate continuity of care, but can increase knowledge transfer and supported choices for women between planned cesarean delivery and vaginal or physiological birth.

Building respect and equality between medicine and midwifery will require investing in communication and building relationships. In order to evolve toward collaborative care teams, maternity care providers must prioritize time to simply sit and talk, with open ears and mutual respect, not only within collaborative hospital teams, but between the providers who work in hospitals and the providers who serve women to give birth at home, in birth centers, or in rural clinics. The quality of communication and trust between these professionals determines whether critical collaborative care moments go well or go poorly, and in an obstetric emergency, this can mean the difference between life and death. Integrated maternity care systems promote mutual respect between doctors and midwives as collaborative professionals, which enables the transparency and continuity of care necessary to optimizing safety and quality of care.[65]

The need for integration based on equality is urgent, as perinatal studies for out-of-hospital birth have made clear that the systems that fail to respect the human right expressed in *Ternovszky* do not prevent the perinatal deaths associated with home birth; they cause them. In the maternity care systems that respect and integrate midwives and uphold women's human rights in childbirth, planned home birth is as safe as planned hospital birth.[66] In the systems where medicine still has a monopoly hold over childbirth and has failed to integrate midwifery, and the state enforces the status quo instead of women's human rights, the result is preventable perinatal mortality for out-of-hospital birth.[67] The only ethical and professionally responsible conclusion to draw from studies showing higher perinatal deaths in non-integrated systems is to work to improve integration, not to work to drive out-of-hospital birth and midwifery further underground.[68] Marginalizing midwifery and out-of-hospital birth as illegitimate, underground healthcare choices is unethical and irresponsible as a policy or a practice.[69] It is well known that it doesn't work to tell women that they are "not allowed" to make personal reproductive healthcare choices.[70] Women make their reproductive choices for reasons that are unique to their own circumstances and history, whether the "authorities" like those choices or not. When the state supports medical monopoly over childbirth by driving midwives underground or refusing to recognize out-of-hospital birth, or when states subject out-of-hospital midwives to unique legal persecution, the result is

- A lack of transparency between doctors, midwives, and their clients during prenatal care,
- A lack of transparency and communication between midwives and their backup professionals during labor,
- Emergency medical services that lack the training and equipment for home birth transfers, but don't respect the midwives enough to let them help,
- Communication breakdowns between midwives and hospital staff during transfer, and even
- Punitive neglect of the birthing women who transfer in from midwifery care,

all of which predictably lead to perinatal, and even maternal, deaths that could have been prevented with timely medical treatment. Because violating *Ternovszky* causes perinatal death, rather than preventing it, no state government can legitimately claim that it optimizes public health by failing to integrate out-of-hospital birth. As an attorney working on cases involving home births, I have seen a pattern of cases in which preventable deaths occurred as a result of these factors of non-integrated care. In each case, the blame and fallout for these dysfunctions was laid squarely on the shoulders of the midwives. Focusing blame on the midwife diverts attention from the system dysfunctions and power dynamics that cause the bad outcomes. Systemic issues are sidelined, and the career and skills of another experienced provider are laid to waste.

It is time to put an end to the hierarchical, monopolistic maternity care systems that were constructed on socially and economically discriminatory systems of sex, race, caste, and of colonial oppression. These systems, and the millions of unnecessary surgeries currently imposed on birthing women, do not optimize maternal and newborn health. Integrated systems with strong midwifery professions as the first line for maternity care, in partnership and with reliable backup from medical professionals, are the most effective, and most efficient, strategy for optimizing the health of mothers, babies, and communities.[71] Healthcare systems are shifting from the old hierarchical models toward "team-based," "patient-centered" care, which is often called "woman-centered care" in the maternity context.[72] This shift reflects a movement away from the vertical model of care, in which doctors were at the top, and everybody else (including the patient) was below them, to a model in which the patient is at the center of a team of care providers, who are working on a horizontal plane, as equal and complementary partners, to provide care as needed for each individual. This transformation is achievable, but only with recognition of the role of power and entrenched inequality in the construction and dynamics of the current system.

Two Maternity Care Paradigms: U.S. and The Netherlands

There is more than one way to understand childbirth, to approach it as a healthcare event, and to experience it as a birthing woman. I discovered this in 2007 when I moved from the United States to the Netherlands, when I was pregnant with my first child. The Dutch maternity care system developed on the model that childbirth is a normal physiological event, with the potential to become pathological or to require medical treatment in some cases. This paradigmatic concept of childbirth contrasted with the model of childbirth on which U.S. maternity care was developed. U.S. healthcare has long framed pregnancy and childbirth as medical events by definition, in which pathology is anticipated and can only be safely avoided or treated through medical management and delivery.[73] The Dutch concept of childbirth as a normal physiological event gave rise to a strikingly different twentieth century maternity care system, in which healthy women gave birth with midwives, usually at home, and were referred to doctors and hospitals only in the event that a medical issue arose during the pregnancy or labor. The Dutch system therefore maintained a strong, independent, well-trained midwifery profession, that has always worked in collaborative partnership with medical providers to serve the birthing population. In the United States, the American Medical Association conducted a concerted publicity campaign in the early twentieth century to eliminate midwifery as a profession, and to put all women under the care of doctors, in hospitals, for childbirth.[74] Childbirth was framed as an event that is pathological in nature and that requires medical intervention—including pharmaceuticals, episiotomies, and instrumental deliveries—even in normal labors.[75] While healthy Dutch women were giving birth at home with midwives, and Dutch doctors were using their skills on the cases that required medical expertise, healthy U.S. women were giving birth under general anesthesia or "twilight sleep," were routinely subject to instrumental deliveries, and were separated from their babies while they recovered from anesthesia and regained consciousness.

Despite the success with which Americans were convinced that giving birth at the hospital was itself critical to safe delivery and good outcomes, the Dutch home birth system has always had better birth outcomes than the U.S. hospital birth system.[76] As the U.S. model was exported to every other developed nation as authoritative and modern, the Dutch system has stood as an indictment of the narrative that hospital birth for all women is necessary to public health and safety. When a European study of perinatal outcomes in 2004, Peristat, indicated that the Netherlands ranked near (but not at) the bottom compared to 29 European nations, hospital birth proponents quickly and vocally declared that the home birth system was responsible for these outcomes. Dutch researchers then conducted the largest home birth study of all time, of over half a million births, to determine whether home birth was affecting its perinatal outcomes. The study proved conclusively that planned home birth, in the integrated Dutch birth system, had equal perinatal outcomes compared to planned hospital birth.[77] Further studies indicated that the causes of the Peristat rankings had more to do with differences in the point at gestation when the Dutch system would implement care for extremely premature babies, so that if 24 or 28 weeks were taken as the baseline instead of 22 weeks, the disparities disappeared. The next Peristat study in 2010 showed the Dutch outcomes to have moved toward the middle compared to other European nations. Nevertheless, media articles had so persistently and effectively drawn a link between Dutch perinatal outcomes and its home birth system that the Dutch home birth rate has fallen significantly, and many people both within the Netherlands and abroad have the inaccurate impression that "studies show" that Dutch home birth kills babies.[78]

Until recently, the majority of Dutch women gave birth at home. Since the media campaign following Peristat, the percentage has fallen to somewhere between 15 and 25%, but this number still reflects hundreds of thousands more babies born at home than any other nation. If home birth increased the risk of perinatal death by 3–10 times, as U.S. obstetricians have asserted, the Netherlands would have long been famous for having many thousands more perinatal deaths than any other nation. Discussions of Peristat and its implications, which turn on the outcomes in 2–4 births per 1000, obscure this basic fact.

As a pregnant American woman planning to give birth in the Netherlands, I had a supported healthcare choice to give birth at home or in the hospital. If I chose to give birth at home, my mainstream health insurance company would not only pay for that choice, but would mail me a box of home birth supplies. I could rely on open, respectful communication between my midwife and the medical providers on whom I might have to rely if a medical problem arose. During my prenatal care, my midwife and our medical backup team would share records and plans transparently, and could get on a call and discuss how to collaboratively resolve any issue with my care in order to maximize safety for my baby and myself. I knew that if I needed to transfer to a hospital during labor, the ambulance services would be trained and equipped for assisting a woman in labor or postpartum, or for providing emergency care to a neonate. I didn't have to fear that hospital providers would impose their moral opprobrium for my choice to give birth at home by neglecting me, ignoring me, or mistreating me, or by refusing to work with my midwife. All of these factors, which come down to the willingness of the healthcare system and the medical profession to support and legitimize women's reproductive healthcare choices, are critical to ensuring safety for out-of-hospital birth. Because the Dutch maternal healthcare system had no political or economic investment in protecting a medical monopoly over maternal healthcare, I could safely choose to give birth at home, with a midwife.

Within a system that enabled the healthcare support for a free choice between giving birth at home or in a hospital, with a doctor or with a midwife, I could explore the clinical reasons for each choice. Like most women, my goal for the birth was to maximize my chance for a healthy outcome for my baby and myself. I wanted to give myself the best chance for a normal physiological birth, while ensuring that medical treatment would be available if the need for it arose.

I was privileged to receive consistent prenatal care and enjoy good nutrition during my pregnancy in the U.S. and after my move to the Netherlands, 7 months into my pregnancy. Which of my healthcare options for labor and delivery would maximize my goal for a healthy, safe birth? While some women would feel safer laboring in a hospital, I anticipated that I would personally feel more relaxed in my home environment. In order to attempt an unmedicated birth, I would need to be able to move, vocalize, and behave as my body dictated during the birth. I knew that I would be less inhibited to do those things in my home, with a midwife I knew and trusted, than in a hospital room with on-call nurses and doctors. I wanted to avoid the imposition of EFM and its associated risks, and to minimize my risk of an unnecessary cesarean section. I would not mind having a cesarean section, or any other intervention, if it was necessary for my baby or myself, but I would mind having a cesarean section that I didn't need.

I gave birth at home with the support of my midwife, Laura van Deth, in 2007, and then again in 2010. In both of these births, my midwife offered me individualized support to overcome complications that could easily have resulted in cesareansections in a hospital setting. Both times, I gave birth in the circumstances that I, personally, needed including immersion in warm water that supported my body through the movements of labor and supported my perineum so that I didn't tear. My midwife handed my baby to me even as she supported it coming out of my body, and was peaceful and quiet around the baby, myself, and my husband as we all gazed on each other for the first time. I experienced the oxytocin bliss that I first read about in Ina May Gaskin's books.[79] The undisturbed minutes after the birth created a sacred memory for our family, and a peaceful initiation into life for my baby, and into parenthood for myself and my husband. Although I could have relied on the medical system to support us if we needed to transfer, we didn't need to go to the hospital during the labor or afterward. We got into bed and rested.

After the birth, I received another normal service of Dutch maternity care: a full time post-partum support nurse, in my home, for at least 8 days after the birth, and up to 2 weeks if the birth is complicated by something like a cesarean section or twins. This *kraamverzorgster* is a nurse who monitors the mother's recovery, measuring her fundus, checking her perineum, and charting her temperature, and also monitors the newborn, charting their conditions in a little book that they can share with their doctors and keep as a record of the postpartum period. She makes sure the mother is drinking and eating, grocery shops, cooks, and cleans—not only laundry and keeping things tidy, but disinfecting the toilet and shower and making the bed up fresh every day. The *kraamverzorgster* teaches a first-time mother everything about caring for a newborn, letting me lie in bed and watch her as I rested the day after the birth, and then adding one new thing that I could do myself, like hold the baby while bathing him without dropping him, until the last day I was doing it all, and she was watching me. As she did all these things for me, I wondered what my friends in the U.S. were supposed to do without such thorough postpartum support—or in many cases, any support at all? How are women without postpartum support supposed to learn all the things about newborn care that my *kraamverzorgster* had taught me—watch a YouTube video? Why wasn't post-partum care part of what they could expect from their healthcare system?

The cost of my home birth and 8 days postpartum support in the Dutch birth system—which was fully covered by mainstream health insurance, with no co-pays for our family—was around $7500. The cost of a vaginal hospital birth in the U.S. that left a woman with stitches and no postpartum support was usually at least twice that, and cesarean sections more expensive than that.[80] Maternal healthcare that meets the real physiological needs of women for labor and postpartum support is not an unaffordable luxury. It is a necessity that women have the human right to expect their healthcare system to be equipped to provide.

The radical implication of *Ternovszky* was that it required the state to dismantle legal and systemic inequality between medicine and midwifery, and to restrain medical monopoly over maternity care. Given the centuries that went into constructing the status quo of that monopoly, it is not surprising that this human right is not yet a reality. Similarly, the power dynamics in place between birthing women, providers, and medical institutions that undermine woman-centered care and informed consent and refusal are based on long histories of inequality, and are currently held in place by powerful economic forces. The human rights framework demands that entrenched systems built on human rights violations realign, or even dismantle, if necessary to fulfill the human right. Governments are obliged to move toward more perfect recognition and protection of human rights, even when doing so is inconvenient to prevailing interests. The evolving global awareness of human rights in maternal healthcare holds the potential to direct a new approach toward eliminating existing dysfunctions and optimizing the systems of support in place for pregnant women and babies.

Endnotes

1. World Health Org., *WHO Statement: The prevention and elimination of disrespect and abuse during facility-based childbirth*, 1 (2014); Bohren MA, Vogel JP, Hunter EC, Lutsiv O, Makh SK, Souza JP, et al. (2015) *The Mistreatment of Women during Childbirth in Health Facilities Globally: A Mixed-Methods Systematic Review*. PLoS Med 12(6): e1001847. doi:10.1371/journal.pmed.1001847.
2. Lozano R, Wang H et al. (2011) *Progress towards Millennium Development Goals 4 and 5 on maternal and child mortality: an updated systematic analysis*, The Lancet.
3. Gaskin, IM, *Ina May's Guide to Childbirth*, Bantam (2003).
4. Leeman, L and Plante, L, *Patient Choice Vaginal Delivery?* Annals of Family Medicine, Vo. 4 No. 3 (2006).
5. American Medical Association, *Opinion 8.08—Informed Consent* (June 2006), http://www.ama-assn.org/ama/pub/physician-resources/medical-ethics/code-medical-ethics/opinion808.page?; American Medical Association, *Informed Consent* (Mar. 7, 2005), http://www.leg.state.nv.us/Session/77th2013/Exhibits/Senate/HHS/SHHS1054M.pdf.
6. Brief of Human Rights in Childbirth et al. as *Amicus Curiae* in Support of the Plaintiff Rinat Dray, *Dray v. Staten Island Memorial Hospital*, Supreme Court of New York, Kings County, Index No. 500510/14. http://www.humanrightsinchildbirth.org/wp-content/uploads/2015/03/Amicus-Brief-Jan-2015.pdf.
7. Berg C, Callaghan W Et al., *Pregnancy Related Mortality in the United States, 1998–2005*, Obstetrics & Gynecology, Vol 116, Issue 6, p1302–1309; Blanchette, H, *The Rising Cesarean Delivery Rate in America: What Are the Consequences?*, Obstetrics & Gynecology, Vol 118, Issue 3, p 687–690 (2011).
8. Digitale E, *Too Deeply Attached: The Rise of Placenta Accreta*, Special Report, Stanford Medicine, Fall 2013. http://sm.stanford.edu/archive/stanmed/2013fall/article4.html.
9. **The Farm Midwifery Center: Preliminary Report of 2844 Pregnancies: 1970–2010,** http://www.thefarmmidwives.org/preliminary_statistics.html.
10. Haelle T, *Your Biggest C-Section Risk May Be Your Hospital*, Consumer Reports, April 13 2016.
11. Eugene R. Declercq, et al., *Listening to Mothers III: Report of the Third National U.S. Survey of Women's Childbearing Experiences*, Childbirth Connection, 35 (May 2013) [hereinafter *LtM III*], http://transform.childbirthconnection.org/wp-content/uploads/2013/06/LTM-III_Pregnancy-and-Birth.pdf.
12. Judy Jou et al., *Patient-Perceived Pressure from Clinicians for Labor Induction and Cesarean Delivery: A Population-Based Survey of U.S. Women*, Health Serv. Res. (Sept. 2014).
13. *LtM III, supra* note 10, at 36.
14. Ibid. at 38.
15. American College of Obstetricians and Gynecologists Committee on Ethics, *Committee Opinion No. 664: Refusal of Medically Recommended Treatment During Pregnancy* (2016); American College of Obstetricians and Gynecologists Committee on Ethics, *Committee Opinion No. 321: Maternal Decision Making, Ethics, and the Law* (2005).
16. Jeffrey Klagholz & Albert L. Strunk, *Overview of the 2009 ACOG Survey on Professional Liability*, 16 ACOG Clin. Rev. 13 (2009); Richard Hyer, *ACOG 2009: Liability Fears May be Linked to Rise in Cesarean Rates*, Medscape Medical News (May 20, 2009), http://www.medscape.com/viewarticle/702712.

17. *See, e.g.*, Sakala, *Least Promising*, *supra* note 39, at e15.
18. Catherine Deneux-Tharaux et al., *Postpartum maternal mortality and cesarean delivery*, 108 Obstetrics & Gynecology 541 (2006).
19. *See* Henci Goer, *Do cesareans cause endometriosis? Why case studies and case series are canaries in the mine*. Sci. & Sensibility (May 11, 2009), http://www.scienceandsensibility.org/?p=147; Anne K. Daltveit et al., *Cesarean delivery and subsequent pregnancies*, 111 Obstetrics & Gynecology 1327 (2008).
20. *See* James M. Alexander et al., *Fetal injury associated with cesarean delivery*, 108 Obstetrics & Gynecology 885 (2006); Anne K. Hansen et al., *Risk of respiratory morbidity in term infants delivered by elective caesarean section: Cohort study*, 336 Brit. Med. J. 85 (2008); March of Dimes, *Analysis shows possible link between rise in c-sections and increase in late preterm birth* (Dec. 16, 2008), http://208.74.202.108/24497_25161.asp.; Astrid Sevelsted et al., *Cesarean Section and Chronic Immune Disorders*, Pediatrics (2015).
21. *See, e.g.*, Emmett B. Keeler & Mollyann Brodie, *Economic Incentives in the Choice between Vaginal Delivery and Cesarean Section*, 71 The Milbank Quarterly 365 (1993) (finding that pregnant women with private, fee-for-service insurance have higher C-section rates than those who are covered by staff-model HMOs, uninsured, or publicly insured); Jonathan Gruber & Maria Owings, *Physician Financial Incentives and Cesarean Section Delivery*, 27 RAND J. Econ. 99 (1996) (arguing that the 13.5% fall in fertility over the 1970–1982 period led ob/gyns to substitute from normal childbirth toward a more highly reimbursed alternative, cesarean delivery); H. Shelton Brown, 3rd, *Physician Demand for Leisure: Implications for Cesarean Section Rates*, 15 J. Health Econ. 233 (Apr. 1996); Joanne Spetz et al., *Physician incentives and the timing of cesarean sections: evidence from California*, 39 Med. Care 535 (June 2001); David Dranove & Yasutora Watanabe, *Influence and Deterrence: How Obstetricians Respond to Litigation against Themselves and their Colleagues*, 12 Am. L. & Econ. Rev. 69 (2010) [hereinafter Dranove] (finding a short-lived increase in cesareans following the initiation of a lawsuit against obstetrician or colleagues); Lisa Dubay et al., *The impact of malpractice fears on cesarean section rates*, 18 J. Health Econ. 491 (Aug. 1999) [hereinafter Dubay] (finding that physicians practice defensive medicine in obstetrics, resulting increased cesarean sections).
22. *See, e.g.*, Nathanael Johnson, *For Profit Hospitals Performing More C-Sections*, California Watch (Sept. 11, 2010), http://californiawatch.org/health-and-welfare/profit-hospitals-performing-more-c-sections-4069 ("women are at least 17 percent more likely to have a cesarean section at a for-profit hospital than at one that operates as a non-profit"); Elias Mossialos et al., *An Investigation of Cesarean Sections in Three Greek Hospitals: The Impact of Financial Incentives and Convenience*, 15 Eur. J. Pub. Health 288 (2005) ("[P]hysicians are motivated to perform CS for financial and convenience incentives."); Hannah G. Dahlen et al., *Rates of obstetric intervention and associated perinatal mortality and morbidity among low-risk women giving birth in private and public hospitals in NSW (2000–2008): a linked data population-based cohort study*, 4 BMJ Open e004551 (2014); Piya Hanvoravongchai et al., *Implications of Private Practice in Public Hospitals on the Cesarean Section Rate in Thailand*, 4 Hum. Res. Health Dev. J. (Jan.-Apr., 200-), *available at* http://www.who.int/hrh/en/HRDJ_4_1_02.pdf (concluding that care in a private hospital, which includes higher rates of intervention, associates with higher rates of neonatal morbidity and no evidence of reduction in perinatal mortality); Kristine Hopkins et al., *The impact of payment source and hospital type on rising cesarean section rates in Brazil, 1998 to 2008*, 41 Birth 169 (June 2014) (noting that publicly funded births in public and/or private hospitals reported lower c-section rates than privately financed deliveries in public or private hospitals).
23. Cherniak D and Fisher J, *Explaining obstetric interventionism: Technical skills, common conceptualisations, or collective countertransference?* Women's Studies International Forum 31 (2008) 270–277.
24. *See* Raghad Al-Mufti et al., *Obstetricians' personal choice and mode of delivery*, 347 Lancet 544 (Feb. 24, 1996).

25. Michael C. Klein et al., *Attitudes of the new generation of Canadian obstetricians: how do they differ from their predecessors?*, 38 Birth 129–39 (June 2011).
26. Katy B. Kozhimannil et al., *Cesarean Delivery Rates Vary Tenfold Among US Hospitals: Reducing Variation May Address Quality and Cost Issues*, 32 Health Aff. 527 (Mar. 2013).
27. Katy B. Kozhimannil et al., *Maternal Clinical Diagnoses and Hospital Variation in the Risk of Cesarean Delivery: Analysis of a National US Hospital Discharge Database*, PLOS Medicine (Oct. 21, 2014), http://www.plosmedicine.org/article/info%3Adoi%2F10.1371%2Fjournal.pmed.1001745.
28. Gaskin, IM, *Ina May's Guide to Childbirth*, Bantam (2003).
29. Ibid.
30. Buckley, S, *Hormonal Physiology of Childbearing: Evidence and Implications for Women, Babies and Maternity Care*, Childbirth Connection (2015).
31. Ibid.
32. Gaskin, IM, Ibid. at 26.
33. Ibid. at 26 and 28.
34. Consumer-driven social media Break the Silence campaigns in the U.S. and Italy: https://www.facebook.com/ImprovingBirth/photos/?tab=album&album_id=705655609507930; https://www.facebook.com/bastatacere/?sk=timeline&app_data.
35. Abuya T et al., *Exploring the Prevalence of Disrespect and Abuse during Childbirth in Kenya*, PLOS One April 2015; Chattopadhyay S., *The Horrifying Sights and Sounds From the Labour Room of an Indian Public Hospital*, Quartz India June 11 2015; Freeze R, *Born Free: Unassisted childbirth in North America*, PhD (Doctor of Philosophy) thesis, University of Iowa, 2008. http://ir.uiowa.edu/etd/202.
36. Gaskin, I.M., *Spiritual Midwifery*, Book Club Co., 4th Ed (2002).
37. See e.g. Swain, J. et al., *Maternal brain response to own baby-cry is affected by cesarean section delivery*, Journal of Child Psychology and Psychiatry 49(10) (October 2008).
38. Murphy-Lawless, J, *Reading Birth and Death*, Indiana University Press (1999).
39. Sartwelle, T, *Electronic Fetal Monitoring: A Bridge Too Far,* J Leg Med 2012 Jul;33(3):313–79; Block J., *Pushed: The Painful Truth About Childbirth and Modern Maternity Care*, Da Capo Press (2008).
40. Dekker, R., *Evidence Based Fetal Monitoring*, July 17, 2012, Evidence Based Birth.
41. Ibid.
42. Steve Lash, *Hospitals: $20.6 M Award Could Spur C-Sections*, The Daily Record (Dec. 7, 2014), http://thedailyrecord.com/2014/12/07/hospitals-20-6m-award-could-spur-c-sections/.
43. *Ternovszky v. Hungary*, 67,545/09, European Court of Human Rights, 2010.
44. The Lancet Series on Midwifery, The Lancet, June 23 2014; National Institute for Health and Care Excellence. Intrapartum care for healthy women and babies. London (UK): National Institute for Health and Care Excellence; 2014.
45. Int'l Federation of Gynecology and Obstetrics, Int'l Confederation of Midwives, World Health Organization et al., *Guidelines: Mother-Baby Friendly Birthing Facilities*, Int'l Journal of Gynecol & Obstet, 2015, 128:95–99; *Executive summary: Collaboration in practice: implementing team-based care.* American College of Obstetricians and Gynecologists. Obstet Gynecol 2016; 127:612–7.
Akileswaran C, Hutchison M, "Making Room at the Table. For Obstetrics, Midwifery, and a Culture of Normalcy Within Maternity Care," ACOG, Obstetrics & Gynecology; 2016.
46. See The Lancet Series on Midwifery, Supra at fn44, and also "C-Section Rates "Vary Widely" Across Europe," BBC News, March 9 2015.; Daniel S., "Why Sri Lanka Beats India in Maternal Mortality Ratios," Al Jazeera, March 14 2016.
47. "Greece Ranks First in the World in the Number of Cesarean Births,"GR Reporter, Dec 8 2011; Barros A, Santos I et al., "Patterns of deliveries in a Brazilian birth cohort: Almost universal cesarean sections for the better off," Rev. Saude Publica, 2011 Aug; 45(4): 635–643.
48. *Muižnieks N, "Protect Women's Sexual and Reproductive Health and Rights," The Commissioner's Human Rights Comments, Council of Europe, July 21 2016, http://www.coe.int/en/web/commissioner/-/protect-women-s-sexual-and-reproductive-health-and-rights.*

49. Fisch D, "Separated At Birth: A Historical and Legal Analysis of U.S. Birth Places and Attendants," Regents of the University of Michigan 2012.
50. Ehrenreich B, English D, *Witches, Midwives and Nurses: A History of Women Healers*, The Feminist Press at CUNY, 2nd ed 2002.
51. Heinsohn G, Steiger O, "Witchcraft, Population Catastrophe and Economic Crisis in Renaissance Europe: An Alternative Macroeconomic Explanation," University of Bremen Discussion Paper, 2004.
52. Ibid, p.25.
53. Ibid, p.23, and see also Ehrenreich B, Ibid, at fn 51.
54. Ibid, p.22.
55. Fisch D, "Separated At Birth: A Historical and Legal Analysis of U.S. Birth Places and Attendants," Regents of the University of Michigan 2012.
56. Sadgopal M, "Can Maternity Services Open Up to the Indigenous Traditions of Midwifery?" Economic & Political Weekly Vol. XLIV 16, April 18, 2009.
57. https://awakenmichigan.org/publications/graphic-and-fact-sheet-u-s-midwives-now-you-see-em-now-you-dont/.
58. Wagner M, *Born in the USA: How a Broken Maternity Care System Must Be Fixed To Put Women and Children First*, University of California Press (2008).
59. See International Center for Traditional Childbearing, https://ictcmidwives.org/; Kline W, "Communicating a New Consciousness: Counterculture Print and the Home Birth Movement of the 1970s," Bulletin of the History of Medicine, Vol 89, 2015.
60. Pascucci C, "Why Are We Asking Doctors If Women Should Have Midwives?" http://birthmonopoly.com/midwives/ Dec 19, 2014.
61. Lusero I, "Making the Midwife Impossible: How the Structure of Maternity Care Harms the Practice of Home Birth Midwifery," 35 Women's Rights L. Rep. 406, 2014.
62. http://pushformidwives.nationbuilder.com/cpms_legal_status_by_state.
63. Louwen F, Daviss BA et al., "Does breech delivery in an upright position instead of on the back improve outcomes and avoid cesareans?", Int J Gynecol Obstet 4 Nov 2016 Accepted Author Manuscript. DOI:10.1002/ijgo.12033.
64. http://www.thebusinessofbeingborn.com/.
65. Ternus-Bellamy A, "Sutter Davis Birthing Center Boasts Lowest C-Section Rate in State," The Davis Enterprise, July 10 2016.
66. De Jonge A, Geerts CC, Ven Der Goes BY et al., "Perinatal Mortality and morbidity up to 28 days after birth among 743,070 low-risk planned home and hospital births: A cohort study based on three merged national perinatal databases. BJOG. 2014; 122:720–728; Birthplace in England Collaborative Group, Brocklehurst P, Hardy P, et al. Perinatal and maternal outcomes by planned place of birth for healthy women with low risk pregnancies: The Birthplace in England national prospective cohort study. BMJ. 2011;343:d7400; Jannsen PA, Saxell L, Page LA, et al., "Outcomes of planned home birth with registered midwife versus planned hospital birth with midwife or physician, CMAJ. 2009; 181:377–383.
67. Wax JR, Lucas FL, Lamont M, et al. *Maternal and newborn outcomes in planned home birth vs planned hospital births: a metaanalysis.* Am J Obstet Gynecol 2010;203:243.e1–8.**;** Snowden JM, Tilden E, Snyder J et al. Planned out-of-hospital birth and birth outcomes. N Engl J. Med. 2015;373:2642–2653.
68. Tilden E, Snowden J, Caughey A et al., "Making Out-of-Hospital Birth Safer Requires Systems Change," Medscape, May 19 2016.
69. Chervenak F, McCullough L, *Planned home birth: the professional responsibility response,* Am J Obstet Gynecol, 2013, Volume 208, Issue 1, 31–38.
70. Haddad L, *Unsafe Abortion: Unnecessary Maternal Mortality*, Rev. Obstet Gynecol, 2009 Spring; 2(2): 122–126.; Rosenthal E, *Legal or Not, Abortion Rates Compare*, NY Times Oct 12 2007.
71. Daniel S., "Why Sri Lanka Beats India in Maternal Mortality Ratios," Al Jazeera, March 14 2016.

72. Shah N. A NICE delivery—the cross-Atlantic divide over treatment intensity in childbirth. N Engl J Med 2015; 372:2181–3.
73. See Block, J, *Pushed*, Supra, at fn37.
74. Ibid.
75. Arnold, J, *Joseph Bolivar Delee and the Prophylactic Forceps Operation*, The Unnecesarean, Dec 6 2009.
76. Infant Mortality Per Thousand Live Births in Six Countries, 1960–2008, OECD Health Data 2011; Infant mortality rates (number of infant deaths per 1000 live births) in 2008 in selected OECD member countries. *Source*: OECD Health Data 2010 (April 2011 version) and WHO Global Health Observatory 2011. http://www.mdpi.com/1660-4601/10/6/2296/htm.
77. de Jonge A et al. Perinatal mortality and morbidity in a nationwide cohort of 529,688 low-risk planned home and hospital births. BJOG 2009 10.1111/j.1471–0528.2009.0217.
78. Devries, R and Buitendijk, S., *Science, Safety and Place of Birth: Lessons from the Netherlands*.
79. Ibid fn 26 and Gaskin, IM, *Spiritual Midwifery*, Book Pub Co, 4th Ed (2002).
80. Truven Health Analytics Marketscan Study, "The Cost of Having A Baby in the United States," Truven Health Analytics, January 2013.

The Periviable Cesarean Section: Can a Case Be Made for Expanding Beneficence in Decision-Making?

Tara A. Lynch and Paul Burcher

Periviable preterm birth is a medically complex and emotionally challenging obstetric scenario characterized by large variations in both clinical practice and patient preferences. Some guidance has been provided for obstetric care providers from organizations such as the American College of Obstetricians and Gynecologists (ACOG) and the Society of Maternal Fetal Medicine (SMFM) (Raju et al. 2014). However, clinical application of these recommendations can be difficult. In order to support both patient autonomy and follow best medical practices, potential obstetric interventions require significant shared decision-making between physicians and patients.

In 2015, we published an article in *Ethics in Medicine* justifying the translation of informed assent, non-dissent, and unilateral physician decision-making to cases of periviable birth (Lynch and Burcher 2016). We agreed with Frank Chervenak and Laurence McCullough who argue that aggressive obstetric management is contraindicated in periviable pregnancies where there is only the potential for iatrogenic harm, without any benefit (Chervenak and McCullough 2013a, 2013b). Using the scenario of a cesarean section for a 22-week fetus, they argue that maternal harm from the cesarean section at this gestation is not counterbalanced by improved fetal outcome. In a straightforward sense, to perform a cesarean section would violate the principle of nonmaleficence because the possibility of patient harm is not counterbalanced by some benefit to her or her fetus. Accordingly, cesarean sections before 23 weeks are generally understood to be medically inappropriate. Furthermore, because the request for cesarean section is most often based on unfounded hopes for

T.A. Lynch (✉)
Department of Obstetrics & Gynecology, University of Rochester Medical Center, Rochester, NY, USA
e-mail: Tara_Lynch@URMC.Rochester.edu

P. Burcher
Alden March Bioethics Institute, Department of Obstetrics and Gynecology, Albany Medical College, Albany, NY, USA
e-mail: BurcheP@mail.amc.edu

L. Campo-Engelstein, P. Burcher (eds.), *Reproductive Ethics*,
DOI 10.1007/978-3-319-52630-0_8

improved fetal outcome, there is general consensus that a cesarean section should not be performed, even upon maternal request (Chervenak et al. 2007). But the question we wish to consider is whether maternal benefits could justify a periviable cesarean section in a setting where there is no conceivable fetal benefit. That is, while we acknowledge that, in general, cesarean sections before viability are likely to violate the principle of nonmaleficence, we would like to consider whether there could be exceptions to this grounded in an expansive notion of maternal beneficence that goes beyond the medical indications and risks of a procedure. The discussion will begin with developing the argument for expanding the notion of maternal beneficence using a recent case we encountered. The second section applies this concept to periviable circumstances and addresses the strengths and shortcomings of this argument by analogy.

Elective Cesarean Section at Term for Maternal Benefit

A recent case we encountered of an elective cesarean section at term first raised this question for us. We have changed some details to de-identify the case. A 40-year-old woman with large uterine fibroids requested an elective cesarean section at 39 weeks for a fetus with a known fatal anomaly. The anomaly would be rapidly fatal after birth, and up to half of fetuses with this anomaly do not survive labor. In general, a cesarean section is not recommended for this fetal condition as there is no conceivable fetal benefit to cesarean delivery for these fetuses. This patient had been counseled extensively about her risks of hemorrhage, hysterectomy, and death, as she did not accept transfusion of any blood products for religious reasons. Her risks of a cesarean section were higher because she had multiple fibroids, which increase blood loss, and she was refusing all blood products. It was clear with extensive counseling that the couple was under no illusion that the baby could possibly survive, and they undoubtably recognized the lack of long-term benefit to the fetus by performing a cesarean section.

The patient had articulated that her goal for the pregnancy was to hold a living baby. Even though she understood the prognosis, having the baby die during labor, before she could hold him, was unacceptable to her. At 40 years old, she recognized that she was unlikely to achieve a pregnancy again, and she stated clearly that she was willing to risk her life in order to hold her son before he died. Her husband supported her decision. They had no lack of clarity about any of the medical facts regarding her risks or the prognosis of her baby. After counseling by both maternal-fetal medicine specialists and an ethics consultation, the care team agreed to perform an elective cesarean section at their request.

We struggled with how to frame our understanding of this case, and settled on an expansive sense of beneficence that moves beyond a strict medical model of physical harm and benefit. While this benefit could be couched in psychological language, we prefer to state it more simply: The patient was expressing a strong desire that was realistic and achievable through an intervention we could provide, albeit at significant risk to her. Although we recognize that this could also be framed in terms of patient autonomy, we will frame it as accepting the patient's appraisal of the good and incorporating it into the weighing of benefits and harms associated with the procedure. Starting from patient autonomy only leads back to beneficence because the right to request a procedure is not unlimited. In order for a patient to choose a procedure, and for a physician to accept this choice, the risks and benefits must be favorably balanced. But in the case discussed here, the balance is only favorable if you expand beneficence to include non-medical goals.

The patient underwent a cesarean section without any complications, and was able to hold her baby for several days before he died at home with them. In this case, we believe, performing the cesarean section was ethically permissible, even though it violates a commonly accepted medical guideline in that she was placed at significant medical risk without any countervailing medical benefit. She was able to fully understand and accept the risks associated with this decision, and therefore was making an autonomous and informed choice to have a cesarean section. While these circumstances are not perhaps unique, it is in our experience relatively uncommon for the care team to acknowledge a larger sense of beneficence beyond a strict medical framework. Had this same patient stated that she desired a cesarean section for an unrealistic sense of fetal benefit, we would argue that this is a distinctly different scenario and a cesarean section would not be ethically permissible.

Expanded Maternal Beneficence in Periviability

In order to illustrate what we mean by expanded maternal beneficence, consider a 40-year-old woman at 22 + 0 weeks gestation with a pregnancy conceived through in vitro fertilization with her last embryo. She has had 4 prior pregnancy losses, including an intrapartum demise at 19 weeks. She presents with preterm premature rupture of membranes and fetal malpresentation. During discussion, the patient states that this is her last attempt to have a child and her one desire is to hold the baby alive. There is another large gush of fluid and a cord prolapse is diagnosed. Is it ethically permissible to perform a cesarean section even though there is no fetal benefit and the fetus will likely die intrapartum? If you do a classical cesarean section, there will be an increased risk of blood loss, as well as an increased risk for transfusion, infection, adhesion formation, and hysterectomy. And this will still not assure fetal survival. Still, perhaps there is a justification in this scenario for a cesarean section if, despite fetal physiologic futility, the procedure would promote both maternal autonomy and beneficence.

This new case can be understood as having ethical relevance by analogy to the previous scenario. While periviable decision-making and decision-making about the term fetus with fatal anomalies are different in some respects, both involve a fetus with little or no hope of survival, which has obvious impact on medical and patient decision-making regarding birthing options. We argue that there are certain situations when performing a cesarean section at 22 weeks gestation is ethically permissible, despite current guidelines, if an expanded sense of beneficence is accepted because the balance of beneficence and nonmaleficence becomes favorable.

Most arguments for refusal of cesarean section for a fetus at 22 weeks gestation or less are justified by the concept fetal physiologic futility (Chervenak et al. 2007). In other words, the cesarean section can have no reasonable expectation to result in the hoped for outcome of a live infant. In this line of reasoning, the risk of harm to the mother is not counterbalanced by any benefit. Justifying refusal of cesarean section in these terms reduces the desired outcome to only fetal survival. However, the mother in our case above is not expecting her baby to survive; her desired outcome is to hold her baby before his expected death. By expanding the desired outcome to include maternal benefits chosen by the woman and grounded in her goals and values, the principles of maternal beneficence and autonomy are being upheld without violating the principle of nonmaleficence.

The idea that non-medical maternal benefit may counterbalance the potential risk of physical maternal harm is not widely recognized by physicians, but we are not the first to suggest this possibility. In fact, a similar scenario was presented in a Mayor and White 2015 *Hastings Center Report* case report by Mejebi Mayor and Amini White. The case involved a request for an elective cesarean section for a fetus with confirmed Trisomy 13. The authors asserted that maternal beneficence and autonomy were being supported by the psychological benefit the cesarean section was providing the patient. By performing a cesarean section the mother was able to participate in spiritual and emotional practices that were important to her, and this sufficiently counterbalanced the potential iatrogenic harm of a term planned cesarean section. Preterm birth is associated with a high rate of maternal depression, post-traumatic stress disorder, acute stress disorders and anxiety. Interventions that potentially mitigate the psychological impact of preterm delivery may reduce maternal morbidity (Greene et al. 2015; Jubinville et al. 2012; Misund et al. 2013, and Vigod et al. 2010). This, in turn, can be seen as promoting beneficence and nonmaleficence.

If the patient's life values and goals are supported by realistic emotional and spiritual objectives that can only be achieved by a cesarean section and are not grounded in false hope, and if the risks of the procedure are fully understood and accepted, then the patient is truly making an autonomous decision that can be supported by physicians within this expanded conceptualization of beneficence. The decision to support a request for elective cesarean section includes some judgment regarding whether the request is grounded in the patient's values, whether the expected benefit is in fact achievable, and if achievable, whether it is only achievable by this more invasive method of giving birth (Chervenak et al. 2007).

While the reasoning above provides support for expanding the notion of beneficence as a justification for periviable cesarean section birth in some cases, it does not provide a rationale for cesarean sections for fetal indication below viability. Current guidelines do not recommend a cesarean section for fetal indications at less than 23 weeks gestation. Furthermore, outcome data indicates those infants born at less than 23 weeks gestation have a 5–6% survival with 98–100% having significant morbidities. (Ecker et al. 2016). At this time, a cesarean section performed at 22 weeks gestation in the hope of improved fetal outcome and survival is medically inappropriate. A physician could not claim that the cesarean section would promote the patient's life goals because the benefit sought by the patient, improved fetal outcome, is grounded in an unrealistic hope. Further, a patient's decision cannot be considered autonomous because the patient is not demonstrating understanding of the medical facts. Therefore, this distinctly different clinical scenario should not result in cesarean section, even on maternal request.

Limits of Expanded Beneficence in Periviability

However, even if the patient has a reasonable justification for a cesarean section, such as our example of a patient with her desire to hold a live infant, there are other problems with periviable decision-making that are not present in our term elective cesarean section example. Other authors have analyzed term cesarean section and determined that the low risk of this particular procedure can support elective cesarean section without fetal benefit (Lannon et al. 2015). But periviable cesarean section is different. At this preterm gestation, a classical cesarean section is frequently required. This particular procedure has increased risks of uterine rupture, abnormal placentation, blood loss, longer operative times and longer hospital stays as compared to low transverse uterine incisions (Gyamfi-Bannerman et al. 2012; Lao et al. 1993, and Bakhshi et al. 2010). In fact, even low transverse uterine incisions (the procedure for a term cesarean section) at preterm gestation have increased risks for future uterine rupture as compared to term cesarean sections (Lannon et al. 2015). The risks of a preterm cesarean section are much higher than that of a term cesarean section, and therefore require a more significant and lengthy discussion with the patient to provide adequate informed consent, and perhaps a greater maternal benefit is needed to counterbalance the increased risk.

Another relevant difference between the scenario of the term fetus with fatal anomalies and the periviable preterm delivery is the amount of time that can be afforded to counseling. The patient with the term fetus had months to meet with counselors, physicians, and ethics consultants. Most importantly, she had time to consider all of her options. This allowed for a unique dialogue to occur which permitted shared decision-making and informed consent. Cases of periviable preterm birth do not follow this same timeline. Not only does the length of the gestation (22 weeks vs. 39 weeks) impact this, but also the urgency that often accompanies these clinical situations. For instance, our 22-week preterm example involves a cord prolapse, which is typically an obstetric emergency. In these cases, decisions are made in seconds rather than weeks. In 2013 Kirsten Salmeen and Cynthia Brincat published a retrospective study of unplanned cesarean sections from 32 to 42 weeks and determined that the interval from informed consent to cesarean section was typically only 50 minutes (Salmeen and Brincat 2013). Other studies have demonstrated that 25% of patients who undergo emergency surgery report insufficient time to consider the consent form (Akkad 2006) and cannot recall the risks of the procedure (Odumosu et al. 2012). Preventive ethics is a proposed method of improving this suboptimal situation: Discussions of complications related to cesarean section could be discussed throughout a patient's prenatal care (Chervenak and McCullough 2013a). However, this seems unrealistic for periviable birth. For some academic institutions the cesarean section rate approaches 30% (Nippita et al. 2015) whereas periviable preterm birth has an estimated incidence of 0.03–1.9% (Chauhan and Cande 2013). Providing counseling regarding periviable decision-making to the general population of obstetric patients is not justified given the low incidence of this complication, and the likely anxiety that such counseling would produce. So while preventive ethics is appropriate for providing more adequate informed consent for cesarean section in general, it is not extendable to the complex informed decision-making required in periviable settings.

Furthermore, these patients are often transported to tertiary-care centers miles to hours away from their homes and are meeting care teams that they have never encountered before. In cases of a periviable preterm cesarean section as compared to a term cesarean section, the short initial interaction between a recently transported patient and the accepting physician may not allow for adequate informed consent for an elective cesarean section, particularly if the discussion includes factors relevant to our expanded notion of beneficence: The patient needs to weigh the considerable risks against her own sense of benefit, and the physician needs to assess whether the patient's expected benefits are founded in the medical facts or not. Non-indicated elective surgeries require careful delineation of risks in order to adequately provide informed consent (Burcher et al. 2013). So, for our example of a 22-week fetus with cord prolapse, it is impossible to imagine that a full discussion of the risks of an elective classical cesarean section could be performed in that limited timeline. The situation does not provide the opportunity to meet the minimum necessary threshold for informed consent to ethically justify a periviable elective cesarean section. In contrast to the majority of term laboring cesarean sections, which are medically indicated, a cesarean section at 22 weeks is elective, in the sense that the indication is maternal request, and currently not recognized by professional organizations. Therefore, the decision to proceed with this would require time, time that may be only afforded in rare scenarios of periviable birth.

Periviability will continue to be a challenging obstetric scenario. With advancing medical technology pushing the threshold of fetal survival to earlier and earlier gestations, viability continues to be a moving target. As the medical facts shift, goals that were previously unrealistic may become more realistic. But this will not change the principles required for good decision-making and the need to communicate complex information in often in stressful circumstances with little time. In the setting of periviability, when maternal goals are realistic, and adequate informed consent of the risks of the procedure has been provided, a cesarean section at 22 weeks gestation may be ethically permissible. Expanding the notion of beneficence to include psychological or spiritual benefit, without opening it up to unrealistic appraisals of the medical situation, admits an exception to the guidelines on periviable decision-making.

References

Akkad A (2006) Patients' perceptions of written consent: questionnaire study. BMJ 333(7567):528
Bakhshi T, Landon M, Lai Y, Spong C, Rouse D, Leveno K, Varner M, Caritis S, Meis P, Wapner R, Sorokin Y, Miodovnik M, Carpenter M, Peaceman A, O'sullivan M, Sibai B, Langer O, Thorp J, Mercer B (2010) Maternal and neonatal outcomes of repeat cesarean delivery in women with a prior classical versus low transverse uterine incision. Am J Perinatol 27(10):791–796
Burcher P, Gabriel JL, Campo-Engelstein L, Kiley KC (2013) The case against cesarean delivery on maternal request in labor. Obstet Gynecol 122(3):684–687
Chauhan SP, Cande VA (2013) Periviable births: epidemiology and obstetrical antecedents. Semin Perinatol 37(6):382–388
Chervenak FA, McCullough LB (2013a) Preventive ethics for cesarean delivery: the time has come. Am J Obstet Gynecol 209(3):166–167
Chervenak FA, McCullough LB (2013b) Ethical issues in periviable birth. Semin Perinatol 37(6):422–425
Chervenak FA, McCullough LB, Levene MI (2007) An ethically justified, clinically comprehensive approach to peri-viability: gynaecological, obstetric, perinatal and neonatal dimensions. J Obstet Gynaecol 27(1):3–7
Ecker JL, Kaimal A, Mercer BM, Blackwell SC, deRegnier RA, Farrell RM, Grobman WA, Resnik JL, Sciscione AC (2016) Periviable birth: interim update. Am J Obstet Gynecol 215(2):B2–B12.e1
Greene MM, Rossman B, Patra K, Kratovil AL, Janes JE, Meier PP (2015) Depression, anxiety, and perinatal-specific posttraumatic distress in mothers of very low birth weight infants in the neonatal intensive care unit. J Dev Behav Pediatr 36(5):362–370
Gyamfi-Bannerman C, Gilbert S, Landon MB, Spong CY, Rouse DJ, Varner MW, Caritis SN, Meis PJ, Wapner RJ, Sorokin Y, Carpenter M, Peaceman AM, O'sullivan MJ, Sibai BM, Thorp JM, Ramin SM, Mercer BM (2012) Risk of uterine rupture and placenta accreta with prior uterine surgery outside of the lower segment. Obstet Gynecol 120(6):1332–1337
Jubinville J, Newburn-Cook C, Hegadoren K, Lacaze-Masmonteil T (2012) Symptoms of acute stress disorder in mothers of premature infants. Adv Neonatal Care 12(4):246–253
Lannon SMR, Guthrie KA, Vanderhoeven JP, Gammill HS (2015) Uterine rupture risk after periviable cesarean delivery. Obstet Gynecol 125(5):1095–1100
Lao TT, Halpern SH, Crosby ET, Huh C (1993) Uterine incision and maternal blood loss in preterm caesarean section. Arch Gynecol Obstet 252(3):113–117
Lynch TA, Burcher PB (2016) Periviability: translating informed assent and non-dissent to obstetrics. Ethics Med 32(1):31–38. http://search.proquest.com/docview/1761754607?accountid=13567
Mayor MT, White A (2015) "Lethal" fetal anomalies and elective cesarean. Hastings Cent Rep 45(6):13–14
Misund AR, Nerdrum P, Bråten S, Pripp A, Diseth TH (2013) Long-term risk of mental health problems in women experiencing preterm birth: a longitudinal study of 29 mothers. Ann Gen Psychiatry 12(1):33
Nippita TA, Lee YY, Patterson JA, Ford JB, Morris JM, Nicholl MC, Roberts CL (2015) Variation in hospital caesarean section rates and obstetric outcomes among nulliparae at term: a population-based cohort study. BJOG 122(5):702–711
Odumosu M, Pathak S, Barnet-Lamb E, Akin-Deko O, Joshi V, Selo-Ojeme D (2012) Understanding and recollection of the risks associated with cesarean delivery during the consent process. Int J Gynecol Obstet 118(2):153–155
Raju TN, Mercer BM, Burchfield DJ, Joseph GF (2014) Periviable birth: executive summary of a Joint Workshop by the Eunice Kennedy Shriver National Institute of Child Health and Human

Development, Society for Maternal-Fetal Medicine, American Academy of Pediatrics, and American College of Obstetricians and Gynecologists. J Perinatol 34(5):333–342
Salmeen K, Brincat C (2013) Time from consent to cesarean delivery during labor. Am J Obstet Gynecol 209(3):212.e1–212.e6
Vigod SN, Villegas L, Dennis C-L, Ross LE (2010) Prevalence and risk factors for postpartum depression among women with preterm and low-birth-weight infants: a systematic review. BJOG 117(5):540–550

Part IV
Abortion, Surrogacy, and Circumcision

"Teaching Morality by Teaching Science:" Religiosity and Abortion Regret

Alesha Doan and J. Shoshanna Ehrlich

Introduction

In 2007, in the case of Gonzales v. Carhart (*Carhart*), the United States Supreme Court upheld the federal "Partial-Birth" Abortion Ban Act of 2003, which criminalized the performance of intact dilation and evacuation abortions (D & E's) unless necessary to save the life of the pregnant woman. Disregarding expert testimony that this late second trimester procedure in which the fetus is removed from the uterus intact may be a woman's safest option, since, among other advantages, when compared to an abortion in which the fetus is disarticulated, it reduces "the risk of cervical laceration or uterine perforation" as well as "the risk that fetal parts will remain in the uterus" (*Carhart* 2007: 161) the Court instead focused on the procedure's "disturbing similarity" to the "killing of a newborn infant" and its resulting "power to devalue human life" (*Carhart* 2007: 158, quoting Congressional Finding (14)(L)). Laying the groundwork for its embrace of the abortion regret trope, which, as we argue in this paper, is rooted in a religious conception of women's place in the natural order, the Court indicates that this procedure is inherently incompatible with women's true nature in light of the fact that "respect for human life finds an ultimate expression in the bond of love the mother has for her child" (*Carhart* 2007: 159).

Alesha Doan and J. Shoshanna Ehrlich equally contributed to this work; the authors' names are listed alphabetically.

A. Doan
School of Public Affairs and Administration, Department of Political Science, University of Kansas, Lawrence, KS, USA
e-mail: adoan@ku.edu

J.S. Ehrlich (✉)
Women's and Gender Studies Department, University of Massachusetts Boston, Boston, MA, USA
e-mail: Shoshanna.Ehrlich@umb.edu

L. Campo-Engelstein, P. Burcher (eds.), *Reproductive Ethics*,
DOI 10.1007/978-3-319-52630-0_9

Although expressly acknowledging that it lacked any "reliable data to measure the phenomena," grounded in this maternalist framing of female identity, the Court nonetheless authoritatively asserted it was "unexceptionable to conclude that some women come to regret their choice to abort the infant life they once created and sustained" and that "[s]evere depression and loss of esteem can follow" (*Carhart* 2007: 159). The sole authority the Court cited for this proposition was the *amicus curiae* brief of the conservative Justice Foundation, which includes sworn testimonies from 180 women injured by abortion gathered by Operation Outcry—a self-described ministry of the Foundation.[1]

Starting with Justice Ginsburg in dissent, who lambasted the majority for its paternalistic invocation of "ancient notions about women's place in the family and under the Constitution [that] have long been discredited" as the basis for depriving "women of the right to make an autonomous choice" (*Carhart* 2007: 188), much criticism has been leveled at the Court's reliance on the concept of abortion regret to justify its result. In addition to likewise objecting to the decision's outmoded gender protectionism, commentators have also critiqued the Court's disregard of scientifically reliable research on the mental health consequences of abortion, and its misapprehension of the concept of regret.[2] However, far less has been written about the religious roots of the regret doctrine and the serious concerns that this raises about the Court's embrace of this antiabortion rationale, which has since infiltrated the legal and political discourse in the guise of a legitimate secular justification for limiting women's abortion rights. This paper seeks to fill this gap.

We begin by discussing our methods, which includes 21 interviews and a textual analysis of two primary sources. Drawing upon our interviews with antiabortion activists, whose work is largely animated by their experiential knowledge of abortion's harm, we highlight the infusion of religious beliefs in their outreach work to counsel "abortion-minded" women.[3] We then examine how male antiabortion leaders harnessed this knowledge to craft a strategy designed to end the nation's "covenant with death" and advance the "Christian renewal of our society" (Reardon 1996: 99), thus importing what had been a privatized therapeutic discourse into the political arena. In conclusion, we document the religious roots of the legal strategy to bring the voices of "post-aborted" women to the Supreme Court in order to persuade the Justices that abortion harms women.

---

[1] The brief was co-authored by Allan C. Parker and Richard Clayton Trotter, Executive Director and General Counsel respectively for the Justice Foundation, as attorneys for the amici, who, in addition to the 180 women injured by abortion, includes Sandra Cano, the former "Mary Doe" of Doe v. Bolton (410 U.S. 179 (1973)), the companion case to Roe v. Wade (410 U.S. 113 (1973)). The testimonies of the 180 women are included in Appendix B and are liberally quoted from in the body of the brief.

[2] For a sampling of this literature, see: Madeira (2014), Siegel (2008a), Siegel (2008b), Turner (2008), Guthrie (2008).

[3] The interviews included in this paper were conducted with activists who counsel "abortion-minded" and "post-abortive" women. Eighteen of the interviewees work in CPCs and three work in prolife organizations that engage in political activism in addition to counseling women. Throughout the paper, we refer to interviewees as activists when referring to the group as an undifferentiated whole and as counselors when referring specifically to those working in CPCs.

Methods

The findings presented in this paper are part of a larger project that is based on 50 in-depth interviews conducted with 26 prolife activists and 24 prochoice activists from 2007 to 2012. In this paper, we exclusively focus on the interviews conducted with 21 prolife activists who have counseled "abortion-minded" and "post-abortive" women. Sixteen of the interviewees are crisis pregnancy center (CPC) counselors and the other five interviewees are activists who have counseled women but work in prolife organizations that engage in a more expansive political agenda that includes traditional and nontraditional political activism. The majority of interviews (16) used in this paper were conducted between April 2010 and November 2012; five were conducted in 2007. Each interview lasted an average of 90 minutes.

The initial purposive sample of activists was obtained through an online search to identify grassroots organizations operating at different locations in a Midwestern state that has an active prolife movement. Concentrating on one state provided an opportunity to explore and contextualize participants' relationships with clients in more depth. Participants were recruited via multiple phone calls and emails to their organizations, and interviews were scheduled with activists who expressed interest. Many participants were reluctant to be interviewed by an academic researcher. Consequently, the recruitment process took several months to establish trust with participants. Engaging in observation at activists' organizations and events served a dual purpose of building trust with activists, and providing background information for the research. We initially interviewed ten individuals affiliated with eight different grassroots organizations. After conducting these interviews, we relied on a snowball sample whereby participants suggested 11 additional individuals to interview affiliated with their respective group.

With the exception of one phone interview, the remaining 20 were conducted in person using a semi-structured, open ended format. Participants were asked general questions pertaining to two themes related to personal beliefs and organizational strategies. For example, participants were asked about their belief system, their beliefs about women seeking abortion, and their involvement in the prolife movement. Organizational questions probed participants about the strategies and tactics used by their respective organizations, the horizontal and vertical relationships between their organization and other groups in the movement, and the influence of their particular group, and the movement as a whole, on policy and culture. Follow up questions were asked for clarification or to probe respondents about a topic, feeling or idea they introduced in the course of answering more general questions.

Sixteen of the interviews were conducted at activists' organizations and one occurred over the phone. Four of the participants preferred to be interviewed elsewhere including coffee shops, hotel lobbies, and private offices. The gender and race of respondents is heavily skewed toward female and white. Sixteen participants are women and five are men; two women identified as African American and the remaining 19 participants identified as white. All interviewees identified as Christian, heterosexual, and their ages ranged from 24 to 74 years old, with 50–59 being the median age category.

Pseudonyms are used for all participants, and human subjects' approval was granted for this research. Participants were given a verbal and written description of the project and each participant was required to read and sign a written consent form. Respondents were provided with a copy of the consent form that apprised them of their rights and contained contact information for the researchers and the human subjects committee that approved the research.

All of the interviews were audio recorded and transcribed verbatim. Interview data were analyzed based on repeated close readings of the transcripts using an inductive and interactive approach to the analysis (Maxwell 2012; Thomas 2006). Through this approach, we used a multi-phased open coding format to catalog dominant and frequent themes that emerged from the data. Ultimately, more themes emerged from the data than we present here, where we focus on prolife interviewees primarily working in CPCs. Across the interviews with prolife activists, we noted themes that fit into the abortion regret narrative: Experiential knowledge of abortion regret, religious beliefs, social construction of women and motherhood, and the aggregation of abortion regret experiences. These themes exist as discrete units but activists frequently wove them together in a larger narrative during the interviews. These themes were then used to continually compare across participants to support themes, refine subthemes, and identify additional themes that emerged over the course of the interviews, which we used to inform the writing of the present paper (Maxwell 2012; Thomas 2006).

We used a similar inductive and interactive approach to conduct a textual analysis of two primary sources written by David Reardon, a prominent antiabortion leader. These two original texts, *Aborted Women: Silent No More* (1987) and *Making Abortion Rare: A Healing Strategy for a Divided Nation* (1996), contain Reardon's vision and strategy for engendering political and spiritual success for the antiabortion movement in the twenty-first century. We initially engaged in a close reading of these primary sources to begin to identify the dominant and frequent themes emerging from the texts, which in turn, were used to refine subthemes, identify other concepts, and compare themes within each primary source. We present the emergent themes from the interviews and textual analysis below.

Abortion Regret in the Crisis Pregnancy Centers

CPCs are the foundation of the outreach branch of the antiabortion movement.[4] Religiously oriented, most CPCs are members of large explicitly Christian-based umbrella networks that adhere to Biblical principles and seek to promote "God's Plan for our sexuality" (Heartbeat International 2016; Care Net 2016). Within this devout and deeply feminized space, the staff offers "abortion-minded" women directive counseling as well as free material resources, such as clothing, diapers,

[4] In this paper, we use the terms "antiabortion" and "prolife." The latter is used when referring to interviewees who self-identify themselves as prolife.

and formula, in order to steer them away from abortion. Counselors center their work around protecting women from abortion regret, which, as developed further below, they experience and understand as a genuine condition.

Much of the intensive and time consuming labor provided at CPCs is performed by volunteers who gain compensation through assisting clients—almost exclusively pregnant women—in a more intimate, relational setting, as compared to other more public forms of antiabortion activism, such as protests and sidewalk counseling (Munson 2009). In any event, those working within this branch of the antiabortion movement, whether volunteer or paid, "feel that feminized, relational approaches carried out woman-to-woman represent the best strategies for preventing abortion and converting clients," and they "often explicitly claim legitimacy for their positions based on their gender" (Kelly 2012: 204, 217). Embodying traditional gender norms and qualities that exemplify nurturance, empathy and selflessness, CPC counselors can thus contribute to the underlying activism of the antiabortion movement through empowering and educating clients without compromising their female-centered identities by affiliating with the more male-identified world of antiabortion politics.

Reinforcing both the gendered and the religious nature of this antiabortion space, our interviews showcase two mutually-reinforcing factors that drive the work of CPC counselors—their individual embodied experience of abortion coupled with an abiding belief that abortion is an active interference in God's ordering of creation—a belief that, as developed below, directly informs this embodied knowledge. Fusing experience and belief, these interviews highlight the dualistic nature of this feminized direct service work—namely, the seamless interweaving of evangelical missionizing into the more familiar goal of dissuading abortion-minded women from terminating their pregnancies.

Our interviews revealed that activists learn about abortion regret through two types of experiential knowledge, embodied or empathetic. Embodied knowledge "refers to personal perceptions of bodily experiences and sensations (e.g. pregnancy)," while empathetic knowledge "is derived from close association with others living with a particular experience (e.g. care-giving)" (Boardman 2014: 138). People's belief systems and moral views shape their interpretation of both forms of experiential knowledge and structure the internalization of it (Casey and McGregor 2012; D'Agincourt-Canning 2005; Potter et al. 2008). Largely owing to the incarnate origin of experiential knowledge, individuals often emphatically embrace their experiences as truth and equate it to more generalized knowledge and fact (Borkman 1976). However, both embodied and empathetic experiential knowledge are rooted in a specific experience and are, by nature, incomplete. The unique and limited scope of experiential knowledge also means that experiences of a condition can only ever be partial because they are bound by time and interpretation (Boardman 2014).

Regardless of the subjectivity and boundaries of experiential knowledge, interviewees discussed abortion regret vis-à-vis their experiential knowledge. While 10 activists drew on embodied knowledge and 11 relied on empathetic knowledge, they all shared a common belief that abortion enables women to circumvent and usurp God's natural ordering of the world. They uniformly identify this irreparable transgression as the cause of abortion regret, which represents the symptomatic manifestation of deviating from this maternal norm. Activists' desire to prevent women from suffering abortion regret was a key motivator for them to participate in the prolife movement.

Among the 16 women we interviewed, ten pointed to their intimate embodied knowledge of abortion regret as the catalyst for their participation in CPCs. The intense aftermath of abortion was acutely felt by these women and their firsthand experiences formed the basis for their understanding of abortion as an act that has long-term—and even irrevocable—consequences. As Marsha, the director of a crisis pregnancy center explained:

> A procedure that was supposedly going to be a simple procedure, had over 30 years of lasting impact on my life ... I've had screwed-up relationships because a fear of people finding out who I was ... And so masquerading who I really was, and not ever letting people get to know who I was ... And that's just part of that guilt and the shame that I carried from having had an abortion that really robbed me (Interview 2007).

Throughout the interview, Marsha repeatedly traced the origin of her life struggle with failed relationships and depression to her abortion. She used her experience with abortion regret to fuel her through the exhaustive hours of emotional labor put in at the CPC because her "personal calling" is to use her experience to counsel women "one-on-one" in hopes that they will forego aborting their pregnancy.

Similarly, Abby, who volunteers her time at a CPC, explains how as a teenager, her lack of agency over the decision to abort, amplified her negative experience:

> My dad made me have an abortion when I was 16. I didn't have a choice. And, that was the worst day of my life. I tried to commit suicide afterwards. I turned into a mess afterwards. So that experience ... it haunted me (Interview 2010).

Retrospectively reflecting on her teenage abortion experience, Abby attributes the difficulties she has had in her adult years, particularly with depression, maintaining relationships, and parenting struggles, to the lifelong arc of abortion regret.

Embodied experiential knowledge of abortion regret populated the stories told by multiple counselors like Michelle who succinctly stated, "People don't know unless you have had an abortion. The depth of the psychological impact, the physical impact, the spiritual impact" (Interview 2012). Across interviews, women often referenced their unique position of being able to relate to other "post-abortive" women because they had "walked in the same shoes." Relying on their own experiences, women claimed authority about the validity of abortion regret and used it as a starting point to interact with clients.

For other activists, including all five male interviewees, abortion regret was a phenomenon that they learned about through empathetic experiential knowledge. Richard's understanding of abortion regret has been informed by the empathetic experiential knowledge he has gained working with "post-abortive" women and helping clients:

> It [abortion] does hurt women. In a lot of different respects. It's kind of overwhelming to think about the huge number of women that have been victimized or are now just devastated by the abortion itself. It's just amazing ... the physical ailments they have to deal with, it's

just unbelievable … The ones who didn't have physical [ailments] now have the psychological problems, and man are they severe (Interview 2012).

Richard uses his experiential knowledge to generalize abortion regret to women beyond his clientele, while concurrently constructing a more expansive conceptual umbrella of abortion regret where he folds physical ailments into the battery of problems it causes.

Echoing Richard, Steven recognizes that his information about abortion regret is empathetically-based, remarking, "I don't want to deny someone's emotions and I haven't lived it—I'm a guy." But he continues to explain how he has witnessed abortion regret firsthand through his activism:

> I've seen women later in life really struggle with things and they don't really know why they're struggling with depression, anxiety, and difficulties in their marriage. And maybe perhaps some of that is because they haven't really addressed all those things and forgiven themselves of the guilt that comes with it [abortion] (Interview 2010).

Steven confidently enlists empathetic experiential knowledge as a source of expertise, which he relies upon to identify the symptoms of abortion regret in women who have not yet connected their struggles to the root cause of their ailments.

Across interviews, activists described abortion as a deliberate and traumatic interference in the God's ordering of creation—a theme that, as discussed below, is central to the development of the woman-protective antiabortion platform. Illustrating this disruption, Dr. Feldon, a CPC counselor, used an analogy to describe the grotesque unnaturalness of abortion to illuminate the emotional harm it causes women:

> I think honestly women finding themselves in a crisis pregnancy whether it's anomaly or not … are like a woman with a leg caught in a trap. She would chew her leg off to get out of the trap. In other words, harm herself. And I think that women naturally, and I think families naturally, don't want to abort … they didn't participate in the termination of that child and all that goes with that and the emotions that are tied within that family and that woman forever (Interview 2012).

By likening women to an animal caught in a trap, Dr. Feldon positions abortion as an unnatural act of self-harm akin to cutting off one's own leg—a conceptualization that is consistent with the Christian-based understanding of abortion as an aberrant disruption of God's biologically determined design for women, which threaded through the interviews.

Mirroring this naturalized biological imperative, Margaret describes abortion as antithetical to God's true purpose for women and she believes arguing with opponents over the medical necessity of abortion misses the key issue at heart:

> I would like for them to value a woman so much that they understood that taking a woman against the grain of nature of what she's created, whether you want to believe she's been created by God or the Sun or some form of being it doesn't matter. Women are created to nurture their young and they follow that pattern even through the animal world … It's not about a medical procedure per se, it's about that woman's welfare and what she's created to be, and we're trying to go against the grain of nature, whatever you want to call nature to be (Interview 2012).

Although Margaret is willing to embrace a more ecumenical view of God, she essentializes her belief that abortion is a gross deviation that goes "against the grain" of the natural order ordained by God to everyone. From Margaret's dyadic religious view, a rejection of God's natural ordering equates to a rejection of God.

Reinforcing the interviewees' conception of abortion as disruptive of God's intended design for women, several interviewees referenced an overt Biblical link, describing the decision to abort as a battle between Christ and Satan for the soul of the "abortion-minded" woman.[5] Drawing on their empathetic experiential knowledge, these counselors relayed stories about the internal struggle pregnant women face, framing it as a tug-of-war between following God's more complex plan for a woman's life, which includes unexpected motherhood, versus being routed into Satan's "quick fix" plan to get rid of her problem and restore a woman to her pre-pregnancy lifestyle.

Reflecting back on his experiences counseling pregnant women, CPC director Jared sees the decision-making process weighted in Satan's favor when a woman who is devoid of a strong religious faith confronts an unwanted pregnancy. As stated, "without a relationship with Christ, without spiritual reasons for doing what you do, it's just natural for people to want to take the low road and make it as easy as they can on themselves" (Interview 2007). Here, Jared references the "low road," which is Satan's "quick fix" plan replete with his disingenuous promise to quickly return a woman to her previous life without spiritual consequence.

Other counselors, however, were not as fatalistic as Jared, and believed that severing Satan's pull can be accomplished even among more secular abortion-minded women. Michelle, a prolife activist, tries to replenish women's spiritual deficit by unmasking the true motivation behind abortion. She draws from the passage in Revelations where "the dragon [is] at the womb of the woman" to illuminate why abortion is engineered for Satan's fulfillment:

> This whole prolife movement is spiritual warfare. We know what the Bible says about life … One of the things I teach on sometimes is that Satan hates women—hates women. Because she brings humanity into the world. So the attack on women … is strong and it's powerful because if he can get rid of her, if he can kill babies—you're killing humanity (Interview 2012).

While Michelle and Jared explicitly implicate Satan as the true culprit behind the lure of abortion, all interviewees implicitly view the abortion decision as a form of spiritual warfare in which the soul of the pregnant woman is up for grabs.

Consistent with the fact that activists self-identify as Christian, and assume aborting women are consigned to a life of despair as the direct consequence of having disrupted God's gendered ordering of the universe, their service model is typically Christ-centered:

> We specialize in helping women in unplanned pregnancies. And our purpose is to save lives. And our goal is to save souls. So we try to help women. We talk about holistic health care and taking care of body, mind, and spirit (Interview 2007).

[5] As discussed in the following section, this is a critical theme in the work of David C. Reardon—a prime architect of the "pro-woman/pro-life" strategy.

To actualize their mission of saving women's souls, counseling sessions, which are free to all CPC clients, are typically laced with religious antidotes that are designed to remove the emotional and spiritual "road blocks" that could potentially derail a woman from following God's path in favor of abortion.

Access to material support is not doled out neutrally, but is often tied to specific requirements such as enrollment in "Bible studies, parenting classes, or abstinence seminars, as activists believe that improving clients' lives requires Christian morals and traditional gender roles" (Kelly 2012: 206). During our interviews, counselors, like Marsha, explained some of the conditional aspects of the services they provide:

> Everything we offer is free of charge ... If they want to visit our friendship room—the diapers and formula and the baby kit products—there's a one page Bible study. We will help anybody one time but if they want to continue to receive our services, it is a requirement that they do that one page Bible study (Interview 2007).

Although Marsha's willingness to offer emotional support is indefatigable, she draws a clear line with material support, outlining how clients are expected to comply with the center's religious teaching requirements as a condition of accessing these resources.

At the CPC where women can earn "mommy bucks" to spend on needed items by attending religious classes:

> There's a lot of education classes [we] offer them. They get to earn mommy bucks. [We] don't just give everything for free; we don't just give you free diapers every time you come in. If you come to our classes and you have earned so many points, then you get to go buy diapers or wipes or whatever (Interview 2012).

By threading religious teachings and practices into their work, counselors thus, simultaneously seek to protect women from abortion regret while also empowering them by strengthening their relationship with God. As discussed in the following section, in the 1990s, this woman-centered approach was appropriated by male antiabortion leaders for political purposes, thus transposing what had been an intimate discourse between women into a powerful strategy for limiting women's control over their reproductive bodies (Siegel 2008a, b).

"One Cannot Hurt a Child Without Hurting the Mother:" The Religious Underpinnings of the "Pro-Woman/Pro-Life" Antiabortion Strategy

In 1987, David C. Reardon published *Aborted Women: Silent No More* in which he concluded, based upon his survey of 282 members of Women Exploited by Abortion (WEBA), that abortion is a "deflating experience which involves a shameful yielding to despair" (Reardon 1987: 315). Undoubtedly anticipating criticism that his findings about abortion's harms were skewed given that, as he openly acknowledged, "WEBA members are women who today regret their abortions," Reardon contended that his respondents were uniquely qualified to speak about abortion regret as they had spent the previous decade coming to grips with the negative impact of abortion on their lives thus giving them a "settled and mature view of their ... experiences" (Reardon 1987: 7–8).

In sharing the kinds of stories that had primarily been cloistered within the feminized spaces of WEBA and the CPCs, Reardon hoped to persuade the public that abortion has two victims—the unborn child *and* its despairing mother. Stressing their conjoined fate, he explained that "[m]other and child are one, and they cannot be separated without doing violence to both" (Reardon 1987: 326). Through this discursive construction of the grieving post-aborted woman, Reardon anticipated the public would recognize it had a moral obligation to end abortion in order to protect women as well as the unborn (Reardon 1987: 320).

However, as Reardon regretfully acknowledged a decade later in *Making Abortion Rare*: *A Healing Strategy for a Divided Nation*, rather than dedicating himself to this public awakening, he had instead found himself having to "explain to pro-life activists exactly *why* post-abortion issues are so important" (Reardon 1996: vii, emphasis in original) to the creation of "a culture in where abortion is not just illegal, but is *unthinkable*" (Reardon 1996: xv, emphasis in original). Extolling the insights of a CPC counselor who confided in him that she had originally been "mostly concerned about the unborn" but after working with "so many young girls who have had abortions" she realized they are the true victims of abortion since "nothing can make a young girl feel more worthless and despicable than having killed her own child," he set about persuading antiabortion activists that "this attitude is the only one which has any hope of creating a pro-life society" (Reardon 1996: 100).

In asking activists to place the post-aborted woman and her narrative of grief at the center of their antiabortion strategy, Reardon was calling upon them to relinquish their singular dedication to the unborn child as their mobilizing icon. As legal scholar Reva Siegel documents, the evocation of "post-aborted" woman's suffering was a strategic move to counter the growing perception that, in the words of Jack Wilke, President of the National Right to Life Committee, "pro-life people were not compassionate to women and that we were only 'fetus lovers' who abandoned the mother after the birth" (Siegel 2008a: 1716, quoting Wilke (2001)). Seeking to reach those who were not convinced that the rights of the fetus should trump those of the pregnant woman, male activists, such as Reardon and Wilke, accordingly "began to experiment with using talk of post-abortion harms ... to persuade Americans outside the ranks of the antiabortion movement that government should impose legal restrictions on women seeking an abortion" for their own protection (Siegel 2008a: 1714). As Siegel writes by strategically "fusing relatively new forms of talk about public health and women's rights with some very old forms of talk about women's roles" they hoped to transform a therapeutic discourse that otherwise, might "have remained embedded in the movement's crises pregnancy centers" into a "woman-protective antiabortion argument" crafted to win over converts to the antiabortion cause (Siegel 2008a: 1714).

In seeking to develop a "pro-woman/pro-life platform" in which the aborting woman would be reimagined from one who is "selfish and immoral" to one who is "confused and despairing" (Reardon 1996: xiv), Reardon found himself facing a double burden of persuasion. In addition to having to convince the "ambivalent majority" that ending abortion was good for women, he faced the daunting task of assuring antiabortionists that this shift would not dilute the movement's moral commitment to the unborn. It is in examining Reardon's strategies for reaching these divergent audiences that the inherent religiosity of the abortion regret doctrine is starkly revealed.

A "Fuller and More Complete" Expression of the "Pro-Life Moral Imperative"

Seeking to persuade potential critics in the antiabortion movement that the adoption of a pro-woman message would not undermine the "moral high ground of opposing abortion simply because all life is sacred" (Reardon 1996: 3), Reardon asserted that "the pro-woman approach is not only consistent with the pro-life moral imperative, it is, in fact, a fuller and more complete expression of it" (Reardon 1996: 4). Tracking the religious views of the CPC activists we interviewed, he explains that: "We begin with a very simple observation. In God's ordering of creation it is only the mother who can nurture her unborn child. All that the rest of us can do, then is to nurture the mother" (Reardon 1996: 4). Grounded in God's dictate that "the interests of the child and the mother are always joined," he thus insists that "from a natural law perspective, we can know in advance that abortion is inherently harmful to women. It is simply impossible to rip a child from the womb of his mother without tearing out a part of the woman herself" (Reardon 1996: 5). Driving home this message, he stresses that "when we are talking about the psychological complications of abortion, we are implicitly talking about the physical and behavioral symptoms of a *moral* problem" (Reardon 1996: 10, emphasis added).

Elucidating why the abortion decision resides in the moral domain, Reardon, in keeping with some of the interviewees, depicts the decision-making process as a pitched battle between Christ and Satan over the fate of a woman's unborn child. Pulling the woman in one direction, Christ urges her not to "do this thing," and implores her to "[p]lace your trust in Me" (Reardon 1996: 108). Pulling her in the opposite direction, Satan insists "[y]ou must get rid of it … You have no choice … Do this one thing and then you will be back in the driver's seat of life" (Reardon 1996: 108).

But all is not necessarily lost for the "desperate woman" who rejects God's gift of life and instead follows Satan to the abortionist's door. If she subsequently repents and embraces his gift of forgiveness, so God may use her as "an instrument for showing the abundant glory of his mercy," she will "escape from the tar pit of despair" in which she would otherwise be trapped (Reardon 1996: 110–111).

However, if a post-aborted woman is paralyzed by the "horror of [her] sin," (Reardon 1996: 11) and thus not believing she is deserving of God's mercy, she will instead find herself consigned to a living hell where Satan seeks to "pump as much despair into [her life] as he can generate" (Reardon 1996: 109). Standing now as her "fiercest accuser," he taunts that she is "beyond redemption … There is no one who can love YOU—a murderer. You are alone," and entreats her to escape this misery by seeking "what little comfort you can in the bottom of a booze bottle, in the silence of suicide, or in the embrace of an affair" (Reardon 1996: 108–109).

This "devil's bargain" by which Satan first encourages a woman to abort and then fans the flames of despair, is aimed at separating women from God—unmoored from her faith, the unrepentant post-aborted woman spirals towards atheism, which Reardon identifies as the "greatest tragedy of abortion" (Reardon 1996: 109). Tracking Satan's jeering admonition that her only hope for comfort lies in death, adultery, or addiction, Reardon likewise asserts that "annihilation of the self," either through the literal act of suicide or through "death's semblance in abusive relationships or the mind deadening effects of drug or alcohol abuse," is her only chance for escape from a life of despair (Reardon 1996: 109, 112).

The concept of the "devil's bargain" crystalizes the animating religiosity of the Reardon's "pro-woman" antiabortion strategy. The tragic figure of the wounded post-aborted woman is the literal embodiment of Satan's victory over God, and her despair the direct consequence of having repudiated his sacred design for her life. So constructed, antiabortion activists are offered the mantle of avenging angel come to wrest suffering womanhood from Satan's vicious grasp.

Seeking to reassure those antiabortionists who might nonetheless be concerned that the adoption of a "pro-woman message" would dilute the movement's moral center, Reardon stressed that grieving mothers are the best spokespersons for the sacred humanity of their unborn children. As he explains, it will be through listening to the "testimony of women who grieve over their lost children," that the ambivalent majority will inexorably be "drawn into implicitly acknowledging the unborn for whom the tears are wept" (Reardon 1996: 9).[6] He thus proclaimed that "by focusing on women's rights, we are not ignoring the unborn, but, instead, are preparing the stage for the most compelling advocates of all for the unborn—their mothers" (Reardon 1996: 14).

Converting the "Ambivalent Majority"

To appreciate Reardon's strategy for converting the "ambivalent majority," it is helpful to understand who comprises this group. Broadly speaking, his typology divides the population into three major groups. First are the "consistently prolife" Christians. Comprising about 33% of the population, this cohort understands that all

[6] As discussed in the final section of this paper, this form of witnessing made its way to the Supreme Court by way of the Justice Foundation's *amicus* brief.

children are a gift from God. Accordingly, the traditional fetal-focused arguments firmly cleave them to the antiabortion cause (Reardon 1996: 18).

The "proabortionists," who comprise about 16% of the population, are at the other end of the spectrum. As Reardon explains, most are "population controllers" who seek to create a utopic society through social engineering, while others are "abortion profiteers" who make their living off of abortions. Deeply invested in abortion as a valued end unto itself, this group cannot be reached even by the message that abortion harms women (Reardon 1996: 18, 29–30).

The remaining 51% of the population are the all-important "ambivalent majority." To varying degrees, those within this category recognize that abortion is the immoral destruction of human life. However, their uneasiness about this child-murder is offset by their concern for the wellbeing of women and a commitment to ideals of female autonomy, resulting in an "uneasy acceptance" of abortion. Consequently, this cohort has hardened its heart "to any moral appeals on behalf of the unborn," (Reardon 1996: 28, 31) thus making it futile to continue to rub "their noses into a truth they already know" (Reardon 1996: 9). Instead, the key to unraveling their "conflicted hearts" lies in persuading them that abortion harms women (Reardon 1996: 25–27). In short, the creation of a prolife society requires battling the opposition on their "*own* turf" (Reardon 1996: X, emphasis added).

However, in contrast to true Christians "who rightly anticipate … that any advantage gained through violation of the moral law is always temporary [and] will invariably be supplanted by alienation and suffering," this constituency does not intuitively grasp that post-abortion despair results from the violation of a moral truth (Reardon 1996: 11). Since appeals grounded in this truth will not be effective in persuading them of abortion's harms, conversion to the cause requires an "alternative way of evangelizing"—namely through the presentation of moral truths as scientific fact (Reardon 1996: 11).

Accordingly, since it is self-evident to the prolife community that moral breaches lead "not to happiness and freedom, but to sorrow and enslavement," Reardon recommends that the development of a research agenda that will produce "compelling evidence" of abortion's harms (Reardon 1996: 11). By then presenting this essential truth as the reasoned outcome of a secular research agenda, antiabortion leaders can effectively "bear witness to the protective good of God's law in a way that even unbelievers must respect" (Reardon 1996: 11). As a result, those whose compassion for the unborn has been blunted by their concern for women will inevitably open their hardened hearts to the sacredness of their humanity.

Significantly, Reardon makes clear that this outreach plan is not simply a blueprint for political reform, but is also a strategy for the Christian renewal of the nation. He prophesizes that evidence of abortion's harms will gradually lead nonbelievers to "appreciate the wisdom of God's laws," leading them to recognize that "maybe all those religious folks weren't so crazy after all" (Reardon 1996: 11). Through the incorporation of post-abortion issues into a comprehensive plan for ending abortion, Reardon thus promises a synergistic campaign "for the defense of human dignity and for the glory of God" (Reardon 1996: viii).

Bringing the Word to the Supreme Court

Reardon's proposed pro-woman plan for ending abortion included a targeted legal strategy, namely the empowering of abortion victims to sue their abortionists for malpractice with the goal of shutting down the industry based on its exploitative and coercive nature. However, in 2000, Allan Parker, President of the conservative Justice Foundation—a self-described "born again Christian who trusts in Jesus Christ as his savior" (http://www.operationoutcrystories.org/2014/08/11/army-of-justice-video) and signs his e-mails "Advancing Life, Liberty, and Justice in Him," (Parker, Christmas Giving, 2015)—was inspired by God to pursue a far grander legal reform plan.

As Parker recounts, in 2000, as he was on his way home from the March for Life rally in Washington D.C., the Lord spoke to him in the Dallas-Fort Worth airport to let him know that "only through the testimonies of women hurt by abortion could [they] refute the lie that abortion is good for women" (Parker 2015). The Lord subsequently instructed Parker to take these testimonies to the Supreme Court in order to persuade the Justices, who, like Reardon's ambivalent majority, had been deceived into thinking that abortion helped women "that you [can]not take the life of your child without it deeply impacting your soul, your body, your emotions" (Parker 2015; Operation Outcry 2014).

The Lord also provided Parker with scriptures to confirm the importance of bringing the testimonies of aborted women before the Court. Significantly, he included a passage from Isaiah, which Parker notes has long sustained the work of the Justice Foundation and its Operation Outcry ministry, predicting that "Hail shall sweep away the refuge of lies and the waters will overflow the hiding place. Your covenant with death will be annulled" (Parker 2015). Although the brief is silent about this revelatory history, the sworn testimonies of the 180 women injured by abortion that reached the Supreme Court in Gonzales v. Carhart by way of the Justice Foundation's *amicus* brief are the direct "fruit of that revelation" (Parker 2015).

Throughout the brief, this testimony is framed as new evidence that did not exist when Roe v. Wade was decided. In keeping with Reardon's political strategy of presenting moral truths as scientific facts, the brief positions women's experiential knowledge as equivalent to scientifically derived findings: "The evidence from post-abortive women now shows that abortion is merely a short-term 'solution' with long-term negative physical and psychological consequences" (*Amicus* Brief 2007: 29).

Further underscoring the importance and credibility of experiential knowledge, the brief favorably cites the 2005 Report of the South Dakota Task Force to Study Abortion (Report) which determined that "prior works indicating that abortion is an emotionally benign medical procedure for most women are invalid and little reliance can be placed upon them" (*Amicus* Brief 2007: 20 quoting the Report 2005, pp. 42–43).[7] Having dismissed this body of scientific evidence as unreliable, the

[7] Empirically sound studies challenging the validity of the idea that abortion leads to despair and regret, include: Rocca et al. (2015), Munk-Olson et al. (2011), Major et al. (2009), Charles et al. (2008), Lee (2003).

brief then relies upon the Report for the wholly unsubstantiated proposition that "it is simply unrealistic to expect that a pregnant mother is capable of being involved in the termination of the life of her own child without risk of suffering trauma and distress. To do so is beyond the normal, natural and healthy capability of a woman whose natural instincts are to protect and nurture her child" (*Amicus* Brief 2007: 19, quoting the Report 2005: 47–48).

The brief accordingly implores the Justices to "re-evaluate" the law and give "deference" to the "real life experiences" of the 180 women injured by abortion as embodied in their sworn testimonies (*Amicus* Brief 2007: 5). This is in direct accord with Reardon's political mobilization strategy to fight abortion vis-à-vis women's embodied experiential knowledge of abortion by bringing the lament of "mothers who mourn" into the public arena.

Consistent with the above-discussed interviews, these testimonies are likewise redolent with themes of post-abortion despair. Significantly, many also contain explicit religious references, such as a post-abortion distancing from God, or a fear of divine retribution, with many others foregrounding religious themes of murder, punishment, forgiveness, and redemption. Unpacking these testimonies thus confirms Reardon's promise to antiabortion leaders that a woman-centric strategy offers a "fuller and more complete expression" of the "prolife moral imperative."

As discussed above, the centerpiece of Reardon's message is founded on a religiously-grounded construction of motherhood that he causally links to the psychological harm created when the inherent bond between mother and child is unnaturally terminated by an abortion. Roxanne Mergenthaler's testimony reflects Reardon's contention that abortion is antithetical to women's true purpose:

> I became psychologically numbed after my first abortion. I was suicidal to begin with then blocked it out and ending up blocking out all of my emotions. I suffer from post-abortion syndrome still, the only healing has been through a bible study, and that has helped. I have two atonement children (not consciously); I've had a hard time bonding with my first child because I got pregnant with her only two months after my abortion (*Amicus* Brief, App. 49).

As Roxanne understands it, the unnaturalness of abortion led her down a destructive, suicidal path resulting from what Reardon denotes as the "psychological complications of abortion" that stem from "a moral problem" (Reardon 1996: 10). Although she sought to atone for the sin of aborting by quickly becoming pregnant again, thus embracing the maternal role she had just rejected, years after her abortion, Roxanne continues to suffer from "post-abortion syndrome," viewing life events through a prism of emotional numbness that is only somewhat improved through bible study.

S.B.M.'s testimony serves as another piece of experiential evidence which supports Reardon's conclusion that "It is simply impossible to rip a child from the womb of his mother without tearing out a part of the woman herself" (Reardon 1996: 5):

> It is with me EVERYDAY—almost 20 years later! For years, I was in denial, but I was bound by shame and guilt. It is the unspeakable deed and harms a woman deep to her core—As a woman, nurturer, child of God … it distorts the image of my life (*Amicus* Brief, App. 50).

S.B.M.'s testimony reifies Reardon's assertion that abortion decimates God's ordained purpose for a woman, and deviating from this maternal norm, leaves behind a long wake of misery that wounds a woman "deep to her core." She traces her emotional regret to the abortion she obtained nearly 20 years ago, connecting her shame and guilt to the trauma created by doing the "unspeakable deed," thus permanently tainting "the image" of her life as a "woman, nurturer, child of God."

Reardon's portrayal of the post-aborted woman who is so mired in despair and deeply ashamed of her sinful behavior that she does not believe she is deserving of God's forgiveness appears in a number of the testimonies. Although the religiosity of Donna M. Razin's testimony was diluted in the excerpted portion contained in the body of the brief, her full testimony captures her highly freighted relationship with God in the wake of her abortion:

> Deep regret—initially I was suicidal—as the years have progressed I have developed a heightened level of bitterness and anger and self-hate. I feared God, have not been able to attend church because of my fear of God, forgiveness, shame, guilt, condemnation, inability to bond and fit in with other women, inability to be intimate. The deep emotional scars were a large contributing factor in my divorce—a very, very catastrophic choice! Great sense of loss and grief (Amicus Brief 2007, App. 12).

As predicted by Reardon, Donna's inability to seek God's forgiveness has delivered her to a living hell that is landscaped with perpetual cycles of loss, condemnation, and self-hate.

Similar to Donna, J.L.M.'s uncut testimony reflects Reardon's supposition that despair and shame will be visited upon the aborting woman, causing her to repudiate God's mercy while simultaneously believing she does not deserve his forgiveness. J.L.M. describes her downward spiral after her abortion leading her to dwell in Reardon's metaphorical "tar pit of despair":

> 11 years later I am obviously still affected. Initially, I suffered from depression, alcohol use increased, increased promiscuity, due to my lowered self-esteem. My grades suffered in college. Relationships were difficult. I had nightmares, flashbacks, and grief. Now with a 6 year old son, I am overly protective to a fault. His relationship with his father is damaged because of my own fears of losing my son. I feel God could still punish me by taking this child away. It's mired my motivation and hindered my career (ironically since my reasoning in part to have an abortion was so my career wouldn't be hindered). It has cut the soul out of my entire life (Amicus Brief 2007, App. 14).

J.L.M.'s testimony reveals that she has not been able to extradite herself from the "tar pit of despair." Particularly poignant, she remains riddled with perpetual anxiety that God's punishing arm may still wound her through her son.

Although both Donna and J.L.M. continues to dwell in this liminal space, where they have lost faith God's plan for their lives, Reardon's promise that the post-aborted woman need not be deposed of hope if she repents of her sin, is illuminated by Cathy's testimony:

> Depression, low-self-esteem, guilt, condemnation, and shame, sleepless nights, nightmares and torment, thoughts of self-hate and suicide, lost, confused, destroyed relationships

> throughout my life, unloved, unlovable, unable to trust God or anyone. Only God can heal this! Post-abortion ectopic pregnancy, fear of not being able to conceive and birth healthy babies, pain, unable to receive love and healing from God and others, worthlessness, strive and drive to perform to be loved and accepted, isolation—emotionally. Lost, confused, needy—shunned by others. I reach out to for help. In a state where I could not forgive myself. Psychologically damaged for the rest of your life (until God heals) (*Amicus Brief* 2007, App. 13–14).

Every facet of Cathy's life has been tarnished and damaged by her abortion regret. She has suffered physical and emotional ailments, and has been plagued with unhealthy interpersonal relationships. However, Cathy's testimony lends credence to Reardon's promise that renewing a spiritual relationship with God is the pathway out of self-flagellating despair.

In accordance with Reardon's expansive articulation of abortion regret laid out in *Making Abortion Rare*, the sworn testimony of post-abortive women included in the Justice Foundation's brief showcases women's lives as broken, mired in addiction, struggling with mental health problems, and riddled with an internal struggle where they see themselves as "beyond redemption" for having aborted their pregnancies. The certitude of women's experiential knowledge has been aggregated and generalized to all women in order to persuade the Court that abortion harms women in furtherance of a two decade old strategy in which antiabortion leaders have marshaled women's experiences to rival and subvert evidence-based research debunking the validity of the claim that abortion harms women.

Conclusion

Tracking the language of Isaiah, Allan Parker explains that standing alone, the testimony of an individual woman about the pain of abortion is but a "hailstone" that "frightens the enemy, but … cannot destroy the refuge of lies that protect abortion in America," but that "a hailstorm, thousands and thousands [of such testimonies] can destroy the refuge of lies" (Parker 2015). This divinely inspired strategy of bringing a hailstorm of testimonies before the Court in order to persuade the Justices that women are harmed by abortion clearly paid off in Gonzales v. Carhart, as they referenced this experiential knowledge as the basis for their conclusion that some women "come to regret their choice to abort the infant life they once created and sustained." Significantly, however, in this regard, the Court relied upon excerpts from the testimonies that had been carefully edited for inclusion in the body of the *amicus* brief so as to avoid any trace of the religious motifs that populate the complete testimonies in the brief's Appendix.

In short, the fruits of Parker's divinely inspired revelation were packaged as secular evidentiary truths regarding the traumatic impact of abortion on women, thus encoding Reardon's strategy to "teach morality by teaching science" into the Supreme Court's abortion jurisprudence.

References

Amicus Curiae Brief of Sandra Cano, The Former "Mary Doe" of Doe V. Bolton, and 180 Women Injured by Abortion as Amici Curiae in Support of Petitioner, Filed with the Supreme Court of the United States in Gonzales v. Carhart No. 05-380 (2007)
Boardman FK (2014) Knowledge is power? The role of experiential knowledge in genetically 'risky' reproductive decisions. Sociol Health Illn 36(1):137–150
Borkman T (1976) Experiential knowledge: a new concept for the analysis of self-help groups. Soc Serv Rev 50(3):445–456
Care Net (2016) About Care Net. https://www.care-net.org/about
Casey L, McGregor H (2012) A critical examination of experiential knowledge in illicit substance use research and policy. J Addict Res Ther 3(5):1000140–1000149
Charles VE, Polis CB, Sridhara SK, Blum R (2008) Abortion and long-term mental health outcomes: a systematic review of the evidence. Contraception 78(6):436–450
D'Agincourt-Canning L (2005) The effect of experiential knowledge on construction of risk perception in hereditary breast/ovarian cancer. J Genet Couns 14(1):55–56
Gonzales v. Carhart (2007) 550 U.S. 124
Guthrie C (2008) Carhart, constitutional rights, and the psychology of regret. South Calif Law Rev 81:877–903
Heartbeat International (2016) Our commitment. https://www.heartbeatinternational.org/about/our-commitment
Kelly K (2012) In the name of the mother: renegotiating conservative women's authority in the crisis pregnancy center movement. Signs 38(1):203–230
Lee E (2003) Abortion, motherhood, and mental health: medicalizing reproduction in the United States and Great Britain. Aldine de Gruyter, New York
Madeira JL (2014) Abortion emotions: regret, relationality, and regulation. Mich Law Rev 21:1–66
Major B, Appelbaum M, Beckman L, Dutton MA, Russo NF, West C (2009) Abortion and mental health: evaluating the evidence. Am Psychol 64(9):863–890
Maxwell J (2012) Qualitative research design: an interactive approach. Sage Publications, Thousand Oaks
Munk-Olsen T, Laursen TM, Pedersen CB, Lidegaard O, Mortensen PB (2011) Induced first-trimester abortion and risk of mental disorder. N Engl J Med 364(4):332–339
Munson ZW (2009) The making of pro-life activists. University of Chicago Press, Chicago
News for Supreme Court and Operation Outcry (25 Dec 2015) E-mail from Allan Parker (on file with the authors)
Operation Outcry (2014) Who we are: http://www.operationoutcrystories.org/about/who-we-are/. Accessed 6 June 2016
Partial Birth Abortion Act of 2003, 18 U.S. C. sec. 1531, Congressional Finding (14)(L)
Potter BK, O'Reilly N, Etchegary H, Howley H, Graham ID, Walker M et al (2008) Exploring informed choice in the context of prenatal testing: findings from a qualitative study. Health Expect 11(4):355–365
Reardon DC (1987) Aborted women: silent no more. Loyola University Press, Chicago
Reardon DC (1996) Making abortion rare: a healing strategy for a divided nation. Acorn Books, Springfield
Report of the South Dakota Task Force to Study Abortion (2005)
Rocca CH, Kimport K, Roberts SCM, Gould H, Neuhaus J, Foster DG (2015) Decision rightness and emotional responses to abortion in the United States: a longitudinal study. PLoS One 10(7). http://dx.doi.org/10.1371/journal.pone.0128832
Siegel R (2008a) Dignity and the politics of protection: abortion restrictions under Casey/Carhart. Yale Law J 117:1694–1800

Siegel R (2008b) The right's reasons: constitutional conflict and the spread of woman protective antiabortion argument. Duke Law J 57:1641–1691

Thomas DR (2006) General inductive approach for analyzing qualitative evaluation data. Am J Eval 27(2):237–246

Turner R (2008) Gonzales v. Carhart and the court's "Women's Regret" rationale. Wake For Law Rev 43:1–43

Wilke J (Feb 2001) Life issues institute is celebrating ten years with a new home. http://www.lifeissues.org/?connector/?01feb.htm

My Child, Your Womb, Our Contract: The Failure of Contract Law to Protect Parties in Gestational Surrogacy

Claire Horner

Melissa Cook, a 47 year old woman who had served as a surrogate once before for another couple, agreed to serve as a gestational surrogate for C.M., the intended father (Langford 2016). Three embryos were transferred, and Cook, who was not genetically related to the embryos, became pregnant with triplets (Cook v. Harding et al. 2016: 7). During the pregnancy, C.M. requested Cook undergo multifetal reduction to reduce the number of fetuses to two to maximize health outcomes for the surviving fetuses and to reduce costs associated with raising triplets (Cook v. Harding et al. 2016: 7–8). Cook, who is reportedly prolife, refused to undergo reduction, and instead offered to adopt one of the resulting children, an offer C.M. adamantly refused (O'Reilly 2016; Cook v. Harding et al. 2016: 8). As a result, both Cook and C.M. filed lawsuits to assert parental rights over the resulting children.

Sherri Shepherd and her husband, Lamar Sally, entered into a surrogacy contract with a gestational surrogate using Sally's sperm and an anonymous egg donor (In re: Baby S. 2015). After the surrogate became pregnant, the relationship between Shepherd and Sally deteriorated, and the parties filed for divorce (In re: Baby S. 2015: 6–7). Subsequently, she also disavowed any maternal connection or parental responsibility for the resulting child, which included a refusal to cooperate in filing the requisite pre-birth parentage paperwork (Crockin 2015). At birth, because the appropriate documents were not filed to determine who would be listed as parents on the birth certificate, the surrogate was named the mother, Sally took custody of the child, and the surrogate was contacted by the state for potential liability for child support (In re: Baby S. 2015: 8).

In the subsequent legal battle in Pennsylvania, Shepherd argued that the surrogacy contract she signed should be void as against public policy (In re: Baby S. 2015: 11). She argued that it attempted to create a parent-child relationship outside the adoption process, and therefore should be void and she should be relieved of her

C. Horner (✉)
Alden March Bioethics Institute, Albany Medical College, Albany, NY, USA
e-mail: claireihorner@gmail.com

L. Campo-Engelstein, P. Burcher (eds.), *Reproductive Ethics*,
DOI 10.1007/978-3-319-52630-0_10

obligations to the resulting child under the surrogacy contract (In re: Baby S. 2015: 11). Although the court disagreed and found Shepherd was the legal mother, and therefore owed child support, the case may have been decided differently if it had occurred in another state—leaving surrogates and intended parents unsure about their rights, responsibilities, and protections under the current contract law framework for gestational surrogacy arrangements.

Conflicts such as these are relatively rare in the context of a commercial surrogacy arrangement, but the nature of the transaction is such that when they occur, they are nearly unresolvable. In states where surrogacy is permitted, principles of contract law are applied to protect the interests of the parties involved. The nature of the reproductive arrangement, however, makes the application of a contract law paradigm inappropriate and insufficient to achieve this goal of protection. Instead, parties to a surrogacy contract are left with lack of certainty and lack of control when conflicts arise, which has the effect of failing to honor their reproductive interests and devaluing all parties involved.

The Surrogacy Arrangement

Surrogacy

Surrogacy is the means by which a woman agrees to gestate a child for the benefit of another individual or couple. As part of this arrangement she also agrees to relinquish any parental rights she may have or to refrain from making a claim for custody or parental rights over the resulting child. Surrogacy has existed in some form throughout human history, but the advent of assisted reproductive technologies such as artificial insemination and *in vitro* fertilization (IVF) have changed how surrogacy is carried out and raised new issues related to its regulation and structure.[1]

Surrogacy involves two parties: The surrogate, who carries the fetus, gives birth to the child, and gives custody of the child to the intended parent(s),[2] and the intended parents, who are the commissioners of the reproductive arrangement and the ones by whom the child will be raised. There are two forms of surrogacy: Traditional surrogacy and gestational surrogacy. In traditional surrogacy, the surrogate not only carries and gives birth to the child, but also contributes her own egg and is artificially inseminated by another, typically the intended father in the reproductive arrangement. Traditional surrogacy therefore involves a surrogate who is both the gestational and genetic mother of the resulting child. Because of increased

[1] For example, traditional surrogacy was described in the Bible. "Now Sar'ai, Abram's wife, bore him no children. She had an Egyptian maid whose name was Hagar; and Sar'ai said to Abram, 'Behold now, the Lord has prevented me from bearing children; go in to my maid; it may be that I shall obtain children by her'" (Gen 16:1–2 RSVCE).

[2] Surrogacy arrangements may involve either one or two intended parents. In this paper, I refer to two intended parents unless otherwise specified.

legal and ethical complexity in traditional surrogacy, this practice has become much less common, and many surrogacy programs offer only gestational surrogacy services (ACOG 2016). In contrast, gestational surrogacy uses *in vitro* fertilization to create an embryo that is not genetically related to the surrogate, which is then transferred into the uterus of the surrogate for her to gestate.

Surrogacy may be further categorized as altruistic or commercial. In altruistic or noncommercial surrogacy, the surrogate is not paid for her services, but only reimbursed for reasonable medical costs, while a commercial gestational surrogacy arrangement provides for compensation for the surrogate. The legality of surrogacy in the United States, in any form, varies by state. Some states presume that all surrogacy contracts are valid and enforceable, some states prohibit surrogacy contracts completely, and other states may allow altruistic surrogacy but prohibit payment to surrogates under the contract (McMahon 2011). While there are several legal and ethical issues raised by the various types of surrogacy, this paper will focus on commercial gestational surrogacy contracts and the unique challenges presented by this kind of arrangement.

The Contract

"A contract is a promise or a set of promises ... the performance of which the law in some way recognizes as a duty" (Restatement (2nd) 1981). It is an exchange of mutual promises that also requires an exchange of consideration (value) for the performance of a service or provision of goods. In essence, each party to the contract gives something to the other side in exchange for some benefit they bargained for.

The surrogacy arrangement is controlled by a comprehensive contract outlining the various rights and responsibilities of the intended parents and gestational surrogate. In a typical contract for commercial gestational surrogacy, the intended parents agree to pay the surrogate according to a prearranged payment schedule that provides a base pay plus applicable medical and other expenses, such as maternity clothes. They also agree to be responsible for the child at birth, including any and all medical decisions and subsequent legal custody. This includes the promise that they will accept custody of the child despite any disability.

The gestational surrogate agrees in the contract to undergo any and all tests recommended by the physician or requested by the intended parents, including amniocentesis. The surrogate also agrees to waive confidentiality to grant access to the intended parents to medical information related to the pregnancy. The contract often contains restrictions based on common medical advice in pregnancy that may cause complications, such as prohibitions on smoking, drinking, drug use, and eating foods that are unsafe in pregnancy.

As a result of the contract, the intended parents receive the child they have collaborated to bring into the world, which ideally has been carried by a healthy surrogate who has optimized the biological environment for the resulting child.

The surrogate receives not only financial compensation, but also the benefit of having given a precious gift to an individual or couple.

These provisions grant broad decision-making authority to the intended parents and attempt to prevent surrogate behavior-linked birth defects or pregnancy complications. However, surrogacy contracts may also include more onerous requirements intended to maximize the health of the resulting child or satisfy certain lifestyle standards of the intended parents. For example, instead of merely avoiding foods that may contain listeria, the surrogate may agree to abide by a more restrictive diet, such as a vegan diet. Requirements to adhere to a regimen of vitamin supplements or avoid unhealthy foods may also be included.

All of these restrictions and requirements are aimed at optimizing the circumstances of pregnancy and birth in the hopes that the resulting child will be as healthy as possible, a goal that is ideally shared by both parties. However, rather than simply providing guidelines for a healthy pregnancy or encouragement to maximize conditions for an ideal outcome, the surrogacy contract formalizes these guidelines as contractually binding provisions. This means that instead of merely facing chastisement for failing to do everything possible to increase the odds of a healthier outcome, the surrogate may also be penalized financially in a lawsuit.

In some extreme cases, it is possible that contracts may even require the surrogate live with the intended parents to allow for supervision and ensure the surrogate is adhering to their standards. Restrictions such as these are difficult to enforce, particularly without constant monitoring of the surrogate. This kind of "quality control" may be clearly articulated in the contract, but the performance relies on the choices of the gestational surrogate, posing significant legal and ethical challenges when conflict arises.

The Limitations and Consequences of Contracts in Surrogacy

Reproductive Conflict

While the contract provisions governing behavioral restrictions may not be overly onerous for a gestational surrogate, provisions regarding reproductive decision-making add a more complex dimension to the bargained-for rights and responsibilities. For a heterosexual couple who conceives intentionally without the use of assisted reproductive technologies, the partners are both the genetic parents and intended parents of the offspring, and the woman is the gestational parent. In assisted reproduction, however, these parental roles—genetic, gestational, and intended—are separated and may be held by up to six individuals in one reproductive arrangement.[3] Each parental role, while it may be well-defined at the outset of the

[3] There may be two intended parents, one gestational surrogate, a sperm donor, an egg donor, and an additional egg donor in cases where mitochondrial replacement techniques are used.

reproductive project, carries with it certain rights, responsibilities, and even emotional attachments that may conflict with the intended plan.

Based on the contract, the intended parents have authority over the fetus; it was their embryo, created and transferred at their direction, and they will be the rearing parents of the resulting child. Although the fetus is physically within the surrogate, the contract grants fetal decision-making authority to the intended parents. In most surrogacy contracts, the parties agree that the gestational surrogate will have the responsibility to terminate the pregnancy at the request of the intended parents, and will not terminate the pregnancy over the objection of the intended parents unless there is a risk of serious harm or death to the gestational surrogate. This agreement is notwithstanding the fact that a woman has a right to reproductive privacy with regard to the decision whether or not to undergo an abortion. Despite the intention of cooperation, the presence of multiple stakeholders in the reproductive arrangement creates the potential for conflict that may not otherwise arise in traditional reproduction. While a pregnant woman in traditional reproduction has both bodily autonomy and fetal decision-making authority, a gestational surrogate has only bodily autonomy, since authority over the fetus belongs to the intended parents. This separation of authority may be easy to conceptualize in theory, but in practice, the intended parents cannot enforce their decisions affecting the fetus without the consent of the surrogate, because to do so would violate her bodily autonomy. As long as the intended parents and the gestational surrogate agree on fetal decision-making, this is not a problem. The issue arises when the fetal decisions made by the intended parents conflict with what the gestational surrogate is willing to allow to be done to her body. Although the fetus may not belong to her, it is still her body that bears the responsibility of gestation; any decision made affecting the fetus will necessarily also affect the surrogate, and what is done to the surrogate's body may also affect the fetus.

It may be argued at this point that with appropriate surrogate matching services such conflicts may be avoided, as no reputable agency would pair intended parents with a surrogate who did not agree with the kinds of reproductive decisions they would make in certain circumstances, such as fetal anomalies. For example, a fertility clinic offering such services would be ill-advised to match intended parents who would choose to terminate a pregnancy for a genetic anomaly with a gestational surrogate who is opposed to abortion in any circumstance. Further, the intended parents are entering into an arms-length transaction in which the surrogate acknowledges that she has no claim or authority over the resulting child and agrees to acquiesce to the intended parents' procreative decision-making. Assuming the parties used a reputable surrogacy agency and were properly matched, why, then, would discussion of such a conflict matter?

Conflicts arise when there is a difference in reproductive expectations between the parties, even if they share the same general values in reproductive decision-making. Whether or not the parties generally agree on the permissibility of abortion, there may be unanticipated situations where the parties' values diverge. For example, the intended parents and surrogate may agree that termination is acceptable in the event of a diagnosis of trisomy 21, but they may not have discussed what happens if the relationship between the intended parents dissolves, or there is financial difficulty and the intended parents want a termination of an otherwise healthy fetus. The surrogate may feel that such reasons for elective abortion are inappropriate, and not part of the original agreement.

Furthermore, even if all contingencies have been discussed and there is no disagreement between the parties, the nature of pregnancy is such that the pregnant woman and the fetus share a biological bond, even if there is no direct genetic relationship (Gammill et al. 2010: 2706). This biological and physical bond may have the effect of encouraging emotional attachment, potentially complicating the surrogate's decision-making. This is one reason why only women who have previously given birth are appropriately considered as surrogates, so as to prevent unanticipated attachment that may make it difficult for the surrogate to give the baby to the intended parents after birth. While the surrogate may be able to overcome this attachment at birth for the sake of the intended parents, this emotional attachment may be enough to change her feelings about termination upon the intended parents' request. She may feel that she is willing to give up the child after birth, but not willing to terminate, even if she had agreed to termination at the outset.

Commodification

Commercial surrogacy has been criticized for its potential for exploitation of women and commodification of her reproductive faculties (Anderson 1990). The practice makes use of a woman's reproductive capacity for the benefit of another, and from the outset of the arrangement, intentionally severs the biological relationship between a gestational mother and the child she carries. Because of the use of the contract to govern the conception, gestation, and subsequent custody of the child, the resulting child is also in danger of being treated as a commodity.[4] The contract not only covers the gestational services of the surrogate, but also requires relinquishment of parental rights, which many have argued is the equivalent of baby-selling (Anderson 2000).

Contract law generally governs marketplace transactions for commodities, whether they are goods or services. While gestation itself may be a service the surrogate provides, other aspects of the contract, such as procreative decision-making, bodily autonomy, and the resulting child, are not commodities subject to contractual control. A purely free-market model would allow parties to trade goods and services of any kind, if such bargains were voluntary, but as a matter of public policy, there are certain rights that are inalienable and therefore not subject to contractual control (Anderson 2000).

As Elizabeth S. Anderson has argued in the context of traditional surrogacy, allowing the parties' preferences to determine the allocation of their alienable rights

[4] "A practice treats something as a commodity if its production, distribution, or enjoyment is governed by one or more norms distinctive to the market" (Anderson 2000).

between them is a hallmark of market transactions (Anderson 2000). In gestational surrogacy, the parties to the contract treat parental rights as if they were alienable property rights, rather than parental rights under a "best interests of the child" paradigm (Anderson 2000). In custody disputes over born children, parents are unable to bargain away their parental rights and responsibilities by contract, because a court will always look to the best interests of the child at the time of the adjudication. While it may not be in the best interests of the child to force a parent to have custody of a child they refuse to be responsible for, there are other benefits, such as ongoing financial support, to which the child still has a right. "One parent is not allowed to pay another parent to go away, and expect such a voluntarily contracted agreement to be upheld by the courts, because parents do not have the right to alienate their rights over their children at will" (Anderson 2000: 20). Traditional surrogacy involves payment to the surrogate in exchange for her relinquishment of parental rights over her genetic child, which essentially "pay[s her] to go away" and ignores the best interests of the child at the time of determination of custody (Anderson 2000).

Although Anderson makes this argument in the context of traditional surrogacy, because the surrogate is required to relinquish her parental rights as both the biological and gestational mother, it can be extended to include gestational surrogacy as well. Anderson's argument that commercial surrogacy commodifies the child by paying the surrogate to give up her parental rights and refusing to consider the best interests of the child relies on the mother-child relationship, both gestational and genetic (Anderson 2000). She acknowledges that gestational surrogacy may alter this analysis as to commodification of the child since the surrogate would not be the legal mother, and therefore would not be paid to relinquish her parental rights because she would have no legal claim to parental rights at all (Anderson 2000, 26n1).

There are two responses to this.[5] First, as discussed previously, although a gestational surrogate bears no direct genetic relationship to the child, she still maintains a maternal relationship to the child that is biological in nature by virtue of gestation, even if she is not the intended parent. Second, if surrogacy contracts only commodify children where the genetic mother is required by contract to relinquish her parental rights without an assessment of the child's best interests, and not where a gestational surrogate relinquishes rights in favor of the genetic parents, then how should a gestational surrogacy arrangement be evaluated where the intended parents have used both a donor egg and donor sperm? If the intended parents bear no genetic relationship to the child, and the surrogate has merely a gestational relationship to the child, the intended parents do not have a "natural right" to the resulting child by virtue of a genetic relationship. What is it about the relationship between the

[5] In my responses, I am assuming the state in which the surrogacy occurs allows for commercial surrogacy contracts, and does not presume the birth mother is the legal mother. In these so-called "intended parentage states," the surrogate would not have automatic standing to sue for custody based on giving birth if she signed a contract relinquishing such custody rights.

intended parents and the resulting child that gives them a greater claim to parental rights than the surrogate?

The contract governing transfer of custody would not be based on the genetic relationship between parents and child, but on the intent of the parties to predetermine custody without considering the best interests of the child. As Anderson states, "I argue that commercial surrogate contracts objectionably commodify children because they regard parental rights over children not as trusts, to be allocated in the best interests of the child, but as like property rights, to be allocated at the will of the parents" (2000, abstract). While one may mount a strong argument in favor of a presumption of parental rights for the intended parents who are also genetic parents, commercial gestational surrogacy arrangements in which none of the parties have a genetic relationship to the resulting child still involve the intentional determination and allocation of parental rights through the contract. The resulting child is not merely returned to the genetic parents, but in fact has been bargained for by the progenitors of the reproductive project.

Legal [Un]enforceability of Surrogacy Contracts

In general, properly formed contracts are legally enforceable, and if one or both parties breach the contract, there are remedies available. Remedies, which can include both monetary damages[6] and specific performance,[7] are intended to make the parties whole, as if they either had completed the contract or had never entered into the contract at all. These remedies not only provide relief to a party who has lost something through a breach, but also encourage both parties to fulfill their promises under the contract to avoid penalty. However, given the intimate nature of procreation and the fundamental rights to both bodily autonomy and procreative liberty, should a surrogacy contract be enforceable to advance the interests of the parties in this situation? Should intended parents be able to exercise complete autonomy over the gestation and birth of their child via surrogate, and should parties to a surrogacy contract be penalized for breach?

---

[6] More specifically, financial remedies may be in the form of expectation damages, reliance damages, and restitution. Damages may also be defined in the contract as liquidated damages. The differentiation and application of these specific remedies is outside the scope of this paper, and therefore I will only discuss money damages generally.

[7] Specific performance is a remedy in which performance of the contract is compelled. For example, in a contract for the sale of a rare or unique item, such as a Stradivarius, it may be difficult or impossible for the buyer to find an equivalent replacement item, unlike a contract for a mass-produced item. In such cases, a court may determine that the item is of such unique value that the only means of making the buyer whole under the contract is to require that the contract be performed, that is, that the Stradivarius be transferred to the buyer according to the terms of the contract.

Standard of Performance

The standard of performance is the standard by which it is determined whether a contract has been breached, and what remedies are appropriate. The standard of performance sets forth the minimum requirements for what the parties expect to receive under the contract. In a contract for the sale of goods, the standard of performance typically follows the perfect tender rule (Perdue 2011: 286). Under this standard, a party receiving the goods may reject goods that are anything less than what is bargained for, even if the deviation is minor (Perdue 2011: 286). In effect, the party providing the goods is in breach if the product is unacceptable to the buyer. In a service contract, on the other hand, the standard of performance is known as "substantial performance." Because service contracts rely on the actions of an individual, rather than the production of a good, a party will only be in breach of the contract if he or she fails to "substantially perform" under the contract; i.e., did the party sufficiently complete performance under the contract, enough that the essence of the contract was performed and he or she is still deserving of compensation under the agreement?

Neither standard of performance seems appropriate to govern surrogacy contracts. If the perfect tender doctrine is imposed, as it is in a contract for the sale of goods, then the intended parents may be allowed to "reject the final product" for any perceived defect—including not only disability but also personal traits such as eye color. This standard would clearly commodify the resulting child, as any child seen as inadequate may be rejected, and the surrogate may be in breach. Such a result is absurd. Conversely, a contract that requires substantial performance may be difficult to enforce where the goal of the contract is to achieve a live, healthy birth. While a surrogate may be considered to have substantially performed if she successfully carries the pregnancy and gives birth to a healthy baby, but failed to adhere strictly to the dietary and vitamin regimen in the contract, it is not as clear if she has substantially performed if she follows all of the contractual requirements throughout the pregnancy, but refuses a medically recommended cesarean section at birth and the child suffers a birth injury. Does performance under the contract rely only on the successful birth and transfer of custody of the child? Is delivery of an acceptable child to the intended parents really the "essence" of the contract? If so, it would seem that surrogacy contracts are more akin to contracts for the sale of a good—the child.

Remedies

Not only is it unclear what is considered a breach of contract, but the remedies available are either unenforceable or inadequate to protect the interests of the parties. Contracts governing the provision of services are generally valid and enforceable, but in the event of a breach, courts are reluctant to order specific performance under the contract as an appropriate remedy (Perdue 2011: 287). Such enforcement would be tantamount to indentured servitude, which is prohibited under the Thirteenth Amendment of the United States Constitution. Particularly in gestational surrogacy, which requires the woman to provide such a personal and intimate service, it is a violation of an individual's general right of self-determination and fundamental right of privacy to force her to complete a reproductive task, even if she had previously agreed to do so in the contract.

In the absence of compelling performance under the contract, the court may award money damages. If a gestational surrogate refuses to undergo an abortion despite the decision of the intended parents to terminate, she cannot be required to have the abortion, but she may then forego future payments under the contract and in some cases may have to repay money she has already received thus far. The same damages may apply where the surrogate chooses to terminate over the objections of the intended parents; the court cannot prevent the surrogate from having an abortion, assuming it is within appropriate medical limits, but the surrogate could be financially liable for breach of contract.

Remedies for breach of contract are aimed at making the parties to the contract whole, and restore them to the position they would be in had they either completed or not entered into the contract. Do money damages for breach of a reproductive contract actually satisfy the interests of the parties involved? While a contract for the sale of goods would compensate the parties by providing a refund or replacing the defective product, refunding money paid under the surrogacy contract does not compensate for the actual loss that has occurred. Where the intended parents have lost their child to an unwanted abortion, or the surrogate refused to terminate upon their request and they are faced with accepting custody of a child they were not prepared to raise, can a contract law framework adequately protect the parties in a surrogacy contract?

Regardless of an agreement to the contrary, such reproductive decisions cannot be enforced over the objections of the surrogate, and so this conflict cannot be prevented by a well-written contract. Commercial surrogacy contracts may be drafted, and portions of them may be enforceable, but no contract can succeed in chaining a surrogate to the provisions of the contract that require her to cooperate with the procreative decisions made by the intended parents in the same way that a contract for the manufacture and sale of goods can control the manufacturing process. Although we can control production of goods under a contract, as discussed earlier, there are certain inalienable rights that cannot be made alienable by contract (Anderson 2000). One such right is the right of reproductive autonomy.

Whose Reproductive Autonomy?

In addition to the legal conflict with seemingly no satisfactory resolution, there remains a deeper ethical concern in commercial gestational surrogacy: Whose reproductive autonomy is being honored? Gestational surrogacy is one of several assisted reproductive techniques aimed at enhancing reproductive autonomy by providing individuals who want to procreate more reproductive options and the ability to exert more control over the process. Reproductive autonomy, or liberty in reproductive decision-making, has historically centered on bodily autonomy, addressing issues such as forced sterilization, forced abortion, or prohibitions on abortion and other maternal behaviors.[8] Because of the separation of reproductive roles, gestational surrogacy also divides the rights inherent in reproductive autonomy between the surrogate and the intended parents, raising the issue of whether surrogacy contracts in fact honor reproductive autonomy at all.

An individual's right to make reproductive decisions, such as when and how to reproduce as well as maternal-fetal decision-making, is fundamental in law and ethics. In gestational surrogacy, however, maternal decision-making and fetal decision-making are split—it is the child of the intended parents, but the body of the surrogate. The intended parents' goal is the birth of a healthy child, and to achieve this goal, they may consider congenital abnormalities, surrogate behavior or the presence of multiples in their fetal decision-making. The surrogate, on the other hand, has an interest in the impact that the pregnancy has on her body, including treatments she must undergo and any effects on her health. "Where these interests conflict, whose rights are stronger: the intended parents of the child, or the woman carrying it?" (Horner 2016).

To answer this question, another question must be asked: what is the nature of the reproductive autonomy that is protected? If it is the right to be free from unwanted interference in matters of reproduction, it seems to support the priority of the intended parents in reproductive decision-making, as they have chosen to engage a surrogate in their reproductive endeavor. However, in the conflict between intended parents and surrogate over certain matters, such as the right or obligation to have an abortion, protecting the intended parents' right to be free from such interference directly burdens the surrogate's own reproductive autonomy. While the surrogate may not have entered into the contract with the intention of adding to her own family, she is still engaged in a reproductive act—one that relies on the use of her reproductive faculties to achieve the reproductive goals of another. While it may not be *her* reproductive arrangement, she nevertheless maintains control over reproduction that requires the use of her body. Reproductive autonomy protects a party from unwanted interference in reproductive decisions, both procreative (when and how to have children) *and* bodily. Where these rights are embodied in different individuals in the same reproductive project, whose autonomy has priority?

If the intended parents' request for an abortion directly challenges the surrogate's autonomy in refusing the abortion, should the surrogate, by virtue of the agreement she entered into, also have an ethical obligation to subordinate her reproductive autonomy to that of the intended parents? To claim the existence of this ethical duty is to limit her moral agency, and restrict her freedom to make her own moral judgment. For example, would this ethical duty be stronger or weaker depending on the intended parents' reasons for requesting the abortion? The surrogate may have entered into the contract with the intention of cooperating to gestate and give birth to a healthy, well-developed child, one free from congenital abnormalities.

[8] Several Supreme Court cases have ruled on various aspects of reproductive autonomy. See, e.g., Roe v. Wade, 410 U.S. 113 (1973); Griswold v. Connecticut, 381 U.S. 479 (1965).

Regardless of a provision allowing the intended parents to request an abortion for any reason, if the intended parents request the abortion because of financial, social, or other non-medical reasons, is it in keeping with the intention of the contract for the surrogate to refuse the request? If, in fact, she can still deliver a healthy child, is she ethically obligated to acquiesce to the intended parents based on a primacy of reproductive decision-making by the progenitors of the reproductive arrangement? To require her to do so against her wishes is to require her to ignore her own evaluation of her values and obligations to both the intended parents and the fetus.

Clinical Issues

The separation of parental roles and the struggle between the reproductive autonomy of the intended parents and the surrogate not only lead to the possible intractable conflict already discussed, but also raise significant clinical issues in the management of the pregnancy and birth. The legal issues raised by the use of a contract have implications for the clinical setting, where clinicians must manage the appropriate medical care of the patient in this complex relationship. The contract, which is intended to protect the rights of the parties in the assertion of their respective reproductive autonomy, may change the traditional physician-patient dynamic by introducing a third party who has (or purports to have) decision-making authority. Similar to medical treatment of a mature minor with parents who are involved in decision-making, treating a pregnant woman carrying the fetus of another can complicate understanding about who should be making decisions, and who, ultimately, is the patient.

Navigating through the autonomy rights of the surrogate and the parental authority of the intended parents may be difficult for the obstetrician. For example, if a surrogate confides in the physician that she has been engaged in health-related misconduct under the agreement, such as smoking or drinking, but refuses to allow the physician to share this information with the intended parents, may the physician disclose this information to the intended parents anyway? After all, the surrogate herself signed a contract previously agreeing to waive confidentiality. Furthermore, where perinatal decision-making becomes necessary, including birth choices and fetal-saving interventions, health care providers may lack clarity about who has authority to make such decisions. If the surrogate is refusing interventions that would be life-saving for the fetus, it may be difficult for the physician to navigate the competing interests of the surrogate, fetus, and intended parents. While a physician has a duty of beneficence and nonmaleficence to his or her patient, what it means to avoid harm and benefit the patient in this context may be clouded due to the nature of the reproductive relationship of the parties. To whom does the physician owe this duty? Ultimately, although the contract binds the parties to the surrogacy agreement, and the surrogate would not be pregnant but for this arrangement, the physician is still caring for the pregnant patient. Regardless of the parties' agreement to the contrary, the physician owes his or her professional and ethical duties to the surrogate, in the same manner as any other pregnant patient.

Conclusion

While the legal and ethical consensus prioritizes reproductive autonomy and allows individuals to use assisted reproductive technologies in myriad ways to achieve the birth of a child, the application of existing legal frameworks to the unique circumstance of gestational surrogacy is unworkable. The essence of the contract is unclear, key provisions are unenforceable, and the contractual reallocation of parental rights causes not only confusion in medical management of the pregnancy, but also commodifies the resulting child. While surrogacy may be the only hope for those suffering from infertility to have a biologically-related child, the potential real-life consequences of a breach of contract in a commercial surrogacy arrangement are severe: There is either a new life or the absence of a new life that will profoundly affect the lives of the parties in ways that are not compensable under a contract.

Acknowledgement I would like to thank Cassandra Rivais for her valuable comments and discussion in the preparation of this paper. I am also grateful to Jane Jankowski and Jazmine Gabriel for reviewing an earlier draft of this paper.

References

(2015) In Re: Baby S. In Pennsylvania Superior Court reports: PA Super
(2016) Cook v. Harding et al. U.S. Dist. Ct. C.D. Ca
American College of Obstetricians and Gynecologists (ACOG) Committee on Ethics (2016) Family building through gestational surrogacy
American Law Institute (1981) Restatement (second) of contracts
Anderson ES (1990) Is women's labor a commodity? Philos Public Aff 19(1):71–92
Anderson ES (2000) Why commercial surrogate motherhood unethically commodifies women and children: reply to McLachlan and Swales. Health Care Anal 8:19–26
Crockin SL (2015) Shepherd-Sally surrogacy agreement upheld by Pennsylvania Appellate Court. ASRM News. Accessed 14 Sept 2016
Gammill HS, Guthrie KA, Aydelotte TM, Waldorf KMA, Nelson JL (2010) Effect of parity on fetal and maternal microchimerism: interaction of grafts within a host? Blood 116(15): 2706–2712
Horner C (2016) My child, your womb. Bioethics Today. http://www.amc.edu/BioethicsBlog/post.cfm/my-child-your-womb. Accessed 28 Jan 2016
Langford C (2016) After fighting abortion demand, surrogate mom demands custody. Accessed 14 Sept 2016
McMahon BM (2011) The science behind surrogacy: why New York should rethink its surrogacy contracts laws. Alb L J Sci Tech 21(2):359–381
O'Reilly K (2016) When parents and surrogates disagree on abortion. The Atlantic. Accessed 14 Sept 2016
Perdue AL (2011) For love or money: an analysis of the contractual regulation of reproductive surrogacy. J Contemp Health Law Policy 27(2):279–313

Vulvar Nick and Metzitzah b'peh: Punishment or Harm Reduction?

Allan J. Jacobs and Kavita Shah Arora

Introduction

Ritual genital cutting in boys and girls can take many forms. At one end of a spectrum are relatively slight interventions, such as nicking the vulva with a lancet. At the other end of the spectrum are radical procedures like penile subincision and vulvar infibulation. Even procedures with little impact have been criticized, especially in Northern and Western Europe (KNMG 2010; World Health Organization 2014), where there is little tradition either of circumcision or of female genital alteration (FGA). In Western nations FGA on minors is especially disfavored, with even minimal forms comprising a crime in a number of European nations (Leye et al. 2007) and in the United States (18 U.S.C. §116).

We limit our discussion to the vulvar nick (VN) and direct oral aspiration of blood from the penile glans immediately following circumcision (*metzitzah b'peh* or MBP).[1] VN consists of a laceration of the vulva or clitoral prepuce 1–2 mm in length, deep enough to draw a small amount of blood, but insufficient to cause other damage. This minimal form of female genital alteration[2] is commonly performed in

[1] *Metzitzah b'peh* means "suction by mouth" in Hebrew.

[2] We use the term "female genital alteration" rather than other terms such as "mutilation," cutting," and "female circumcision" that are used to name these ritual acts because "alteration" is more

A.J. Jacobs (✉)
Gynecological Oncology, Coney Island Hospital, Brooklyn, NY, USA

Department of Obstetrics and Gynecology, Coney Island Hospital, Brooklyn, NY, USA

Department of Obstetrics and Gynecology, Stony Brook University, Stony Brook, NY, USA
e-mail: allanjoeljacobs@gmail.com

K.S. Arora
Department of Obstetrics and Gynecology, MetroHealth Medical Center, Case Western Reserve University, Cleveland, OH, USA
e-mail: Kavita.Shah.Arora@gmail.com

L. Campo-Engelstein, P. Burcher (eds.), *Reproductive Ethics*,
DOI 10.1007/978-3-319-52630-0_11

Indonesia (Corbett 2008), among other places. MBP is part of the circumcision ritual of some ultra-Orthodox Jews. Unlike circumcision, MBP and all types of FGA are practiced by groups that are marginal in Western nations. We assume, at least for argument's sake, that neither FGA of any sort nor MBP may ethically be performed. It is insufficient, however, merely to condemn a procedure. It is also important to discuss the ethics of possible methods to effectuate a judgment that a procedure is wrong. Measures that promote adoption of ethical norms themselves have ethical implications and consequences.

To that end, we shall maintain that VN and MBP should not be punished in a liberal state. From a utilitarian perspective, the costs of such punishment are likely to outweigh the benefits. From a retributionist perspective, any effective punishment would need to be unjustly disproportionate to be effective. Therefore, punishment should be eschewed, while attempts to maximize safety of traditional practices should be encouraged.

In the remainder of this paper we make several assumptions for purposes of discussion: (1) that both VN and MBP performed on minors are ethically undesirable; (2) that ritual infant male circumcision is legal, as currently is the case throughout the Western world; (3) that ritual genital procedures performed on minors are not intended to provide sexual stimulation or gratification to the infant or to any adult involved, and that they do not do so; (4) that these procedures do not humiliate the recipients in the context of the communities that practice them; and (5) that MBP causes serious infection in one of several thousand boys, while the only injury inherent in VN is a small transient laceration. Acknowledging that this has been a subject of fierce debate, we do not assert here a position regarding the ethical appropriateness of non-therapeutic circumcision, but note that it is a widespread and legal phenomenon, and proceed on the basis of this fact. To the extent that these assumptions are untrue or invalid, our conclusions might be invalid as well.

It is not accidental that we choose to consider VN (as an alternative to more extensive forms of FGA) and MBP together. First, both are practiced by small and marginal groups. Second, the objective harm they induce is minimal (with VN) or rare (with MPB). Finally, both can be part of either a punitive or a harm-reducing regimen by governments or other institutional actors. We first review the medical risks of VN and MBP and then we discuss acceptable government approaches. We conclude that intolerance of MBP and VN leading to punishment creates more harm than good. We instead favor a positive approach that encourages and refocuses efforts at education and promotion of safer alternatives.

Vulvar Nick (VN) as an Alternative to More Extensive FGA

FGA of minors is endemic among various religious and ethnic groups in Africa and also is common in certain Islamic nations in Asia (World Health Organization 2014). Over 100 million living women have undergone some form of FGA. The

---

neutral and accurate than any of the alternatives. For a more in-depth discussion of this nomenclature, please see Arora and Jacobs (2016).

European Institute for Gender Equality (EIGC), an agency of the European Union, has estimated that 500,000 girls and women in Europe have been subjected to FGA, and that 180,000 are at risk each year (EIGE 2013: 25). However, data collection is poor, rendering estimates unreliable (EIGE 2013). EIGC believes that most females in Europe who have been subjected to FGA either immigrated following FGA or are taken to their countries of origin to receive it. A major newspaper has reported that the absolute magnitude of FGA in the United States is similar, citing unpublished data from the Centers for Disease Control and Prevention (Topping 2015). FGA is performed for cultural or religious reasons (Gruenbaum 2005) though the practice may not be mandated by any religion's canonical texts.[3] The likelihood and severity of FGA health risks depend on the extent of the procedure (Obermeyer 1999; Shell-Duncan 2008). The most radical procedure, infibulation, markedly narrows the introitus, resulting in dyspareunia, increased obstetric morbidity, and neonatal mortality (Eke and Nkanginieme 2006).[4] A novel classification system for FGA based on risk of the procedures has recently been proposed as an alternative to the existing WHO classification system (Arora and Jacobs 2016).[5] Under this categorization, VN is a Category 1 procedure, having no foreseen long-term impact on morphology or function. Pain probably is slight, and can be minimized with using local anesthetic cream. Long-term effects are nonexistent.

FGA has been condemned as violating human rights, and activists wish to eliminate all forms of FGA (Shell-Duncan 2008). Some African nations have criminalized

---

[3] It is not useful to try to categorize FGA as being either religious or cultural. Whether or not Islamic or other texts require it, popular belief is not necessarily congruent with canonical interpretation. Any practice with a spiritual basis has religious meaning to its practitioners (Orsi 2003). Also, distinctions among spheres of human activity such as religion, sexuality, culture, and politics that many scholars in the West take for granted (Walzer 1983) are not universally applicable, as the main concern of many religions is *praxis*, which can pervasively affect most life activities. It is also not useful to distinguish between religious and cultural norms and "mere" social pressures, as the former are reinforced by the latter. Consider, for example, the likely effect on your social status if you served at a dinner party a dog barbecued on a spit. Consider, also pressures on new Jewish, Hindu, or Mahayana Buddhist parents to assign names to their children based on traditional constraints.

[4] The study cited is the most extensive, and best designed, to date. However, and contradictory to the title, it is not a prospective study, as patients were not enrolled until they went into labor. Consequently, the study did not optimally control for differences between women who had undergone FGA and those who had not, though attempts were made to compensate for this problem through the statistical design.

[5] The entire classificatory scheme is as follows:

Category 1: Procedures that have no lasting effect on vulvar morphology or function.

Category 2: Procedures that create morphologic changes, but not functional changes.

Category 3: Procedures that are likely to impair the ability of the recipient to engage in or enjoy sexual relations. If a procedure also is likely to impair reproduction function it should be placed in Category 4.

Category 4: Procedures likely to impair reproductive function, either by reducing the chances of conception or by making vaginal delivery more dangerous.

Category 5: Procedures likely to cause major physiological dysfunction outside the vulva, including death, even if performed correctly.

FGA, with various degrees of energy devoted to their enforcement and various degrees of success when enforced (Shell-Duncan et al. 2013). In some instances, cultural change has come through educational efforts (Gillespie and Melching 2010). Emigration from Africa has brought the practice of ritual FGA to Europe (Leye et al. 2006) and North America. All FGA, including VN, is felonious in the United States.

Harborview Hospital in Seattle, Washington encountered a local immigrant Somali population whose members desired hospital infibulation for their daughters (Coleman 1998). After discussion, leaders of the Somali community, and the hospital, and its physicians agreed to perform VN of the clitoral hood instead of the traditional practice of infibulation to assenting girls with the idea that they would not then undergo infibulation. Anti-FGA activists, led by a prominent congresswoman, conducted a campaign against this agreement. Under this pressure, Harborview reversed its decision to perform VN as a harm-reducing procedure to prevent more radical forms of FGA from being performed outside the health care system. Somali women then stated that they would now instead take their daughters to Somalia or to Somali practitioners in the United States for the traditional procedure (Coleman 1998). Under these circumstances, infibulation, rather than VN, would be carried out—often without analgesia or asepsis—resulting in pain, possible psychological trauma, and physical morbidity and possibly mortality.

Years later, the American Academy of Pediatrics (AAP) deemed that a "ritual nick" in the "clitoral skin" (more precisely, the skin of the clitoral prepuce) was an ethically acceptable compromise that "may build trust between hospitals and immigrant communities," may save some girls from undergoing more extensive and damaging procedures, and "play a role in the eventual eradication" of FGA (American Academy of Pediatrics 2010a: 1089). However, the AAP retracted this opinion under pressure and reissued the original language that opposed all forms of FGA including VN (American Academy of Pediatrics 2010b). If government and the larger society equally disfavor infibulation and the VN, communities may be less likely to abandon more harmful procedures for less harmful alternatives (Arora and Jacobs 2016).

Metzitzah b'peh (MBP)

Ritual circumcision of Jewish boys is performed on the eighth day of life by a religious official called a *mohel*. Removal of the blood on the penile glans is termed *metzitzah*. Most *metzitzah* is performed using a gauze sponge or a pipette. Some ultra-Orthodox rabbis require that the *mohel* remove the blood by direct orogenital suction, a process known as *metzitzah b'peh* (MBP) (Kerlee 2004). The procedure is not mandated by other Orthodox rabbis, many of whom believe that it poses an unacceptable threat to an infant's life (Korobkin 2006). MBP is opposed by non-Orthodox Jewish movements (Otterman 2012). MBP may infect the infant with herpes simplex virus (HSV) if performed by an infected *mohel*. Although usually a mere annoyance beyond infancy, HSV can cause serious infections in neonates. The natural history of neonatal HSV is well-characterized because up to 1 in 3000 infants acquire HSV from their mothers through vaginal delivery (Cherpes et al. 2012). Half of HSV infections either are disseminated or involve the central nervous system. Approximately 25% of infected infants with disseminated HSV die even with prompt treatment, but over 75% die if treatment is delayed. Neurologic sequelae follow herpes encephalitis in 15% of infants. The MBP-specific HSV infection rate is about 1 in 4000 (Centers for Disease Control and Prevention 2012).

HSV infections associated with MBP prompted the Department of Health of New York City (which is home to many ultra-orthodox Jews) to require *mohelim*[6] to obtain written parental consent prior to performing MBP. A Federal appellate court ruled that the United States Constitution demanded strict scrutiny, so that the government would have to demonstrate that the regulation was the narrowest possible remedy satisfactory to address a compelling government interest. (*Central Rabbinical Conference v. New York City Department of Health and Mental Hygiene* (763 F.3d 183 [2nd Cir. 2014]).[7] The Board of Health, under a new mayor, has abandoned the regulation (Goldberg 2015).[8]

We are unaware of any law specifically prohibiting MPB. Absent specific intent to inflict harm or to give or receive sexual stimulation, MBP would not constitute sexual abuse. However, the language of some child abuse statutes might be applicable to MBP.[9]

Punishment and Non-punitive Approaches: General Considerations

There are three major theoretical bases for punishment: Utilitarian, retributional, and expressive. Utilitarian punishment is mostly deterrent or rehabilitative in intent (Tadros 2011). Retributional punishment is aimed at assigning to the violator a fate

---

[6] Plural form of mohel.

[7] Strict scrutiny is a constitutional test applied to government actions that impinges on certain liberties or rights the Supreme Court considers fundamental. It requires that the restrictive measure the narrowest possible remedy available to address a compelling government interest. This test does not consider practicality or cost of alternative measures, and is a hard hurdle for the government to overcome.

[8] There is an informal agreement between the city and communities practicing MBP that *mohelim* who transmit HSV through circumcision will be permanently banned from practicing. It is not known whether either the city government or the communities that practice MPB have taken any steps to implement this understanding or to monitor compliance.

[9] An example is the Nebraska Revised Statutes, §28–710: "(2) For purposes of the Child Protection and Family Safety Act: ... (b) Child abuse or neglect means knowingly, intentionally, or negligently causing or permitting a minor child to be: (i) Placed in a situation that endangers his or her life or physical or mental health." Most American states have language applicable to MPB; the Nebraska statute was chosen for quotation because of its breadth and the clarity of its language.

that the violator deserves (Zaibert 2013). Expressive or communicative punishment (Harris et al. 2011; Wringe 2012) is directed toward affirming something that is important to the state; both the violator and the victim are secondary. An example of such a law is prohibition of either the slaughter of horses or the sale of horsemeat for human consumption (Cavel International, Inc. v. Madigan 500 F3d 551[(7th Cir., 2007]). Retributional and expressive punishments are deontic; expressive theories may also be based on virtue ethics.

It is difficult to determine whether a punishment realizes a utilitarian goal because numerous other variables change simultaneously. For example, some states reacted to high crime rates by instituting long prison terms for drug offenses and "three-strike" laws mandating long incarceration for repeat felons. The utilitarian motivation included specific deterrence (preventing criminals from repeating their acts because they are incarcerated) and general deterrence (general fear of prison and fines).[10] The rate of most violent crime has fallen markedly.[11] However, this has been associated with other changes that conceptually might be causative. First, the decline in violent crime is temporally associated with a falling birthrate and fewer unwanted births due to a wider availability of contraception and elective abortion, resulting in a smaller proportion of young adults and perhaps fewer young adults especially prone to commit crimes. There also have been changes in policing procedures and adoption of precautions by potential victims. It therefore is difficult to know what proportion of the decline to attribute to increased incarceration.

With retributive approaches, the punishment presumably is its own justification. Rationales include expiation of guilt, moral education to the community, and satisfying (in part) the wrongs die to a victim. It also emphasizes the idea that citizens are responsible for their behavior. Retributive theories imply a necessity for proportionality. This need not be the one-for-one proportionality stipulated in the *lex taliones* (*see* Exodus 21:23–25). Some American laws, such as the RICO Act (18 U.S.C. 1964), provide for treble damages.[12] Nonetheless, retributive punishment incorporates a relationship between the magnitude of misdeeds and consequences. Expressive punishment also provides its own justification. Expressive punishment is likely to exceed limits imposed by proportionality, though. For example, a Texas law used to impose a year in jail for burning an American flag.[13] Thus, a purely expressive basis for punishment, divorced from just utilitarian or retributive ends, seems to be antithetical to liberal principles.

---

[10] A third possible utilitarian goal is rehabilitation, or providing the offender with the attitudes and tools to lead a lawful life.

[11] From 1992 to 2012 annual incidence of violent crime fell from 1,922,274 to 1,214,464 (36.8%) (Bureau of Justice Statistics, available at http://www.bjs.gov/ucrdata/Search/Crime/State/RunCrimeTrendsInOneVar.cfm).

[12] Damages are explicitly restorative, however. They do not constitute pure punishment in the sense that imprisonment, loss of parental care, or even fines punish people without restoring losses other than some dignitary losses to victims.

[13] The Supreme Court voided this law as unconstitutional (*Texas v. Johnson*, 491 U.S. 397 (1989)).

Utilitarian punishment theory is forward looking, considering the consequences of punishment. Retributive theories, on the other hand, are backward looking. They seek to redress past wrongs, whether on behalf of the victim or society, or against the offender (Rawls 1955). The basis for a criminal offense is utilitarian: To protect society. In contrast, the basis for a specific act of enforcement is generally thought of as retributive. We have laws against murder to discourage murder and thus protect safety, but we punish an individual murderer because he broke the law. We believe that both utilitarian and retributive considerations must be satisfied. Untempered utilitarian theory would allow the framing and punishment of an innocent person to prevent greater harms (for example, an insurrection). But retributive theory would justify vigilante justice in cases in which guilt is clear. We believe that punishment should both satisfy conceptions of justice and be likely to achieve an overall increase in good. Punishment of trivial offenses likely serves neither objective, though.

Utilitarian Punishment, VN, and MBP

Punishing people for VN or MBP is likely to be unsuccessful at deterrence, as discussed below. Furthermore, such punishment may cause more harm than good overall, as punishment harms not only the punished party but those associated with that person, especially their families. Children should not be friendly-fire casualties in policy wars over rites of passage that involve genital alteration. Furthermore, the harms caused by suppression may extend beyond wrongdoers and their families, affecting society as a whole. Under these circumstances, pursuit of perfect conduct with regard to ritual genital procedures is the enemy of the good. Even if unwarranted, some of these procedures should be tolerated, but governments and doctors should attempt to encourage performance of relatively safe procedures over more dangerous ones to protect the health of the children involved.

Governments have a panoply of tools at their disposal that they can employ to combat practices such as VN and MBP. Laws can provide for fines and imprisonment against parents and those who perform the procedures. Also, courts have great leeway in restricting parental authority. They can impose measures as drastic as permanent termination of the parent-child relationship in the best interest of a child, as when harm to a child has occurred or may be imminent (*Guardianship of KHO*, 736 A.2d 1246, [S.C.N.J., 1999]). Not all measures in the state toolbox are punitive. Governments also can regulate procedures, as by requiring consent, licensure, or that the procedure be performed only under certain restrictions.[14] Governments also

[14] For example, the Texas legislature enacted a statute requiring that abortion clinics conform to requirements of ambulatory surgical centers, and that providers have staff privileges in nearby hospitals. This statute has been overturned, however, by the United States Supreme Court (*Whole Women's Health v. Hellerstedt*, No. 15–274; ___ U.S.___ (2016)) as unconstitutionally creating an undue burden on women's reproductive choices. However, the Court previously upheld a statute requiring a 24 hour waiting period for elective termination of pregnancy (*Planned Parenthood v. Casey*, 505 U.S. 833 (1992)).

can implement educational programs or propaganda campaigns. Finally, government can incentivize desired behavior.

The success of any approach depends, in part, on the importance of the practice to the target groups. A multi-pronged campaign in the United States has aimed at smoking. Measures have included high tobacco taxes, restricting smoking in public venues, education, providing resources for those who wish to stop smoking (Zhu et al. 2012), and age limits for purchasing tobacco products (Steinberg and Delnevo 2013). These measures have been associated with decline in the percentage of adult Americans who smoke from 42% in 1965 to 19% in 2012 (Centers for Disease Control and Prevention 2015). Prohibition of alcohol through legislation was less successful. Though it reduced drinking, it was unpopular (Blocker 2006). It therefore was repealed after less than 13 years, even though this required the difficult effort of amending the Constitution (Schrad 2007). Since then, the percentage of adults using alcohol has remained steady at about 60% (Gallup 2015) despite taxation and restrictions. The importance of a practice to those who engage in it, differences in danger arising from the practices, and differences in legitimacy of government in the eyes of potential violators all may influence the ability of government to curb it.

Criminalization carries undesirable externalities. First, it disrupts the lives of those who are punished and of those who are close to them. Second, punishing members of a minority culture for activities it considers appropriate and important can alienate its members. This alienation not only can be counterproductive to reducing the undesirable practice, but can result in conflict, lawlessness, and rebellion. Third, externalities related to law enforcement may actually exacerbate the harms caused by the outlawed practice. For example, a legal regimen designed to suppress the trade of a recreational drug with mildly deleterious health effects might worsen the public health problem if the laws result in illegal trade in adulterated drugs.

Punishment of VN and MPB: Consequences to Individuals Involved

MBP performed by a *mohel* not infected by HSV or other transmittable illnesses poses negligible long-term risk to the individual involved. If *mohelim* can be screened for infection so that *mohelim* that shed HPV do not perform MBP, then harms are maximally reduced. The difference between this and VN is that the policy goal is to reduce the potential for harm from MBP, while VN is, itself, a modality for reducing harm from more extensive forms of FGA. For both VN and MBP, though, the magnitude of the public health problem from the procedure is small (though MBP causes devastating harm to a few individuals). We do not know of a case of criminal conviction or modification of parental rights based on performance of VN or MBP. This is fortunate. Termination of parental care after the procedure would not benefit the children, since their parents would not subject them to further genital trauma. Termination would represent an attack against minority cultural norms. In summary, punitive measures against adult participants in MBP and VN are likely both to be unjustly disproportionate and to create more harm to the individuals involved than they avert. This would be true even if they were partially successful in reducing the incidence of these two procedures.

Consider the effect of imprisonment as a felon for allowing one's child to undergo MBP (from a virus-free *mohel*) or VN. The imprisoned person is confined in an unpleasant and dangerous place. She is unable to earn a living and is separated from her family. After discharge, she loses various rights, which may include the right to vote or to obtain professional licenses. If the crime is characterized as a sex crime, her name and photograph may be posted permanently online, and her residence may be restricted or publicized. The offender's family also suffers. The imprisoned parent is not providing care or income. The children may be stigmatized. If both parents are imprisoned, the child is removed from the family unit, and may be exposed to the vicissitudes of the foster care system (Petersilia 2001). Instability arising from a parent's imprisonment is likely to interfere with the child's socialization and education. Following release of the imprisoned parent, all must adjust to her reintegration into the family, which is often unsuccessful (Naser and Visher 2006). The effect of such punishment on a child's financial, social, emotional and educational circumstances exceeds the negative effect of VN or MBP, which, in any event, will not be repeated.

Punishment of VN and MPB: Consequences to Minority Groups and to General Society

It is unlikely that the ultra-Orthodox Jewish groups that practice MBP can be induced to give up the practice, as they consider it to be integral to their religious practice. They would continue to practice it even at the risk of prosecution. These groups also tend to interpret religious law as forbidding them to collaborate with civil authorities in prosecuting actions that the community believes to be appropriate (Broyde 2002). Furthermore, members would consult their rabbi before reporting anyone to civil authorities (Resnicoff 2012). It is almost inconceivable that they would receive rabbinic permission to report a *mohel*'s performance of MBP. Furthermore, parents afraid of state punishment might delay in bringing an HSV-infected child for medical care, and prompt treatment is important for a good prognosis.

Mechanisms of social cohesion and control among Muslim immigrants that practice FGA are not as well documented. Since FGA is felonious, those who perform it will try to prevent its existence from becoming known outside the community. There is little information on the incidence of FGA and its complications in Western nations. Even in Africa, years of advocacy and legislation aimed at eliminating FGA have seen little decline in the prevalence of the practice. This is a testament to the importance of FGA to those that practice it and to the relative lack legitimacy held by secular authority.

Laws against FGA and MBP are likely to drive the procedures underground. Girls then will undergo extensive genital alteration in lieu of VN. And parents may be reluctant to take boys who acquire HSV infection due to MBP to a physician early in the course of the infection because of fear of punishment. This concern is not entirely speculative. Deleterious public health effects of a punitive regime have been noted in analogous circumstances. For example, prior to legalization of elective termination of pregnancy, severe morbidity and mortality from unsafe abortions were common (One Hundred Professors of Obstetrics and Gynecology 2013). Currently, parents in the Western world desiring FGA for their daughters obtain this despite laws to the contrary, either by taking their daughters to a neighboring country or by using lay practitioners (Shell-Duncan 2001).

There are other utilitarian reasons to avoid enacting unpopular laws. Jürgen Habermas observed that although the threat of coercion must underlie laws, the laws will not be effective unless most people obey them voluntarily (Habermas 1988). Voluntary compliance with law, in turn, is grounded in government legitimacy, "the recognition of power (felt observation to authority and a corresponding duty to obey) [combined with] justification of power (through shared moral purpose in co-production of social order)" (Jackson et al. 2012: 1054). Tyler et al. distinguish between relational legitimacy, based on a perception of procedural fairness (Tyler 2006) and instrumental legitimacy, based on substantive agreement with the organization's decisions (Huo et al. 1996). When people do not so identify with a rule-imposing organization then legitimacy, if present, is instrumentally-based (Huo et al. 1996). To the extent that people are committed to the behavioral demands of a religion they can ascribe only instrumental legitimacy to government, as transcendental religious claims are likely to trump conflicting laws (Jackson et al. 2012). Laws that conflict with divine mandates cannot be obeyed. Efforts to enforce those laws will cause those religious opponents to regard government as a hostile force.

Persecution affects the persecutors as well as the victims. As Erwin Staub said regarding genocide,

> Individuals and groups change as a result of their own actions. Acts that harm others, without restraining forces, bring about changes in the perpetrators, and in other members of the group, that make further and more harmful acts probable. In the course of this evolution, the personality of individuals, social norms, institutions, and culture change in ways that make further and greater violence easier and more likely. (Staub 1999: 182).

This principle is applicable to lesser forms of coercion than threat of death. Persecution of a minority delegitimizes that minority. It also creates a substantial group with a tangible interest in persecution. This group comprises people such as police, prison officials, and attorneys. Attempts to suppress MBP would involve potential action against thousands of parents and *mohelim* annually. As for FGA, Leye cited recent studies documenting thousands of girls in various Western European nations to be at risk for FGA (Leye et al. 2006). These predictions probably underestimate the number of girls currently at risk, as there is continued heavy immigration to Europe and America from nations where FGA is prevalent. In Sweden, with a population of less than 10 million, there are an estimated 38,000 girls and women who have had FGA procedures (7000 under 18) with an additional 7000 girls at risk (Forslind 2015).

Public choice theory suggests that government officials with a stake in a program have an interest in perpetuating and extending it. An example of this phenomenon is support by a union representing California prison guards for a law providing for prolonged incarceration for perpetrators of a third felony, even if non-violent. This predictably led to increased incarceration, with consequent employment opportunities for prison guards (Borland 1994). That is not to say that police want to perpetuate crime or that physicians would promote smoking so that they will have more patients to treat. Rather, when there are multiple credible policy options, those with a vested interest will be inclined to take positions that enhance those vested interests. Their conscious motives need not be consciously self-serving. Rather, they may believe they are doing good and important work. Many California prison guards likely believed that the three-strike laws protect the public.

Another adverse effect on society arises from the theory of cognitive dissonance, which predicts that criminalizing a practice leads to perception that those who perform such a practice are bad people. Adherents of religions that engage in disfavored practices may be marginalized (Bushman 1960). The proposition that criminalization reflects and promotes prejudices against minorities is evidenced by the differential treatment of ethanol and peyote as used in religious rites, and by the differential treatment of circumcision and VN. The directionality of the causation is not clear—whether attempts at suppression lead to disdain for the target or vice versa; these may be mutually reinforcing.

The combination of mutual contempt between majority and minority generated by policies designed to eliminate VN or MBP, combined with the immediate investment of a segment of society in the suppressive program, would harden division between majority and minority groups, make it more difficult for the minorities to compromise, and create an atmosphere of mutual intolerance that detracts from achieving a liberal ideal. When the minority population is large, this can create social disorder.

We conclude that utilitarian considerations weigh against a punitive approach to VN or MBP. We now consider the deontic rationales—retribution or expressiveness.

Deontic Theories of Punishment, VN, and MBP

We have argued that retributional punishment must be proportional by definition. Even if the purpose of the penalty is expressive, there must be some sort of justice principle at work in a liberal democratic state. For example, a long prison term for desecration of a national totem such as a flag or a portrait of a government official seems unjust. Unjustly severe punishment undermines deontic justification for that punishment.

Furthermore, European jurisprudence regards criminalization (the use of punishment) as *ultima ratio* (last resort) (Jareborg 2004). Punishment should be reserved for the worst behavior—for "secular sins" that are "worthy of unconditional condemnation" (Jareborg 2004: 534). Tuori has equated *ultima ratio* with proportionality (Tuori 2013). Bengoetxea sees *ultima ratio* as the mean between "impunity where blameworthy and harmful conduct goes unpunished and overpunity where any undesired conduct is criminal" (Bengoetxea 2013: 113). Criminal sanctions should be used only when private and administrative measures fail (Jareborg 2004), and then should be proportionate to the crime. The same can be said of restriction of parental rights.

What sort of punishment is proportional for a single, minimally painful nick of the vulva that does not cause physical harm or psychological humiliation? Similarly, what sort of punishment is proportional for oral aspiration of blood from the penis in the absence of humiliation or sexual gratification? Removal of the child from the home would seem to be a disproportionate response, as any injury has been minimal. The criminal implication of these acts varies by jurisdiction. They are not sexual offenses, which require demonstration of specific intent.[15] Both VN and MBP might be punishable as less serious crimes.[16] It is highly debatable, though, whether a prison sentence is appropriate for acts that do not result in significant harm, and which are seen as positive measures within the community of those who carry out the acts. Thus, application of either criminal penalties or restriction of parental rights seems grossly disproportionate in the case of either VN or MBP. Thus, there are deontic reasons why punishment of MBP and VN are unwarranted.

Harm Reduction as an Alternative to Punishment

Harm reduction is substitution of a less harmful for a more harmful procedure (Marlatt 1996). Marlatt characterizes harm reduction as a public health model. He contrasts it with the medical disease model and the non-medical moral model, both of which demand abstinence from the harmful behavior. While recognizing abstinence as ideal, supporters of harm reduction encourage any movement toward decreased harm as a step in the right direction (Marlatt and Tapert 1993). Harm reduction seeks buy-in by those who exhibit the target behavior. Needle exchange programs for users of injected recreational drugs and programs designed to encourage moderate alcohol use among underage drinkers are examples (Marlatt and Witkiewitz 2002). If consent procedures and a program of HSV surveillance in

[15] For example, the New York Penal Code requires proof that the act had the purpose of "sexual gratification of either party" (§130.00). Each element of a crime, including intent to produce sexual gratification, must be proven beyond reasonable doubt.

[16] The New York Penal Code defines child endangerment as conduct performed knowingly and "likely to be injurious to the physical, mental or moral welfare of a child" (§ 160.10 (1). It is a Class A misdemeanor, punishable by up to 1 year imprisonment.

mohelim reduce the incidence of HSV infection in circumcised boys, or if girls undergo VN instead of infibulation (Shell-Duncan 2001), disease burden from MBP and FGA would be lower. In fact, the use of VN as a means of reducing the harm of FGA has been thwarted by activists not only in the United States, but in Italy (Galeotti 2007) and the Netherlands (Obiora 1996). To our knowledge, the New York informed consent regulation that was struck down judicially was the only harm reduction program targeting MBP. However, the same ethical considerations apply as in VN; we believe that harm reduction programs are strongly indicated to the extent that it MBP and FGA threaten children's health.

Unfortunately, we do not know if these programs would have been effective because they were not applied. However, a recent report describes the successful adoption in Khuzistan (the southwestern, predominantly Arabic-speaking, province of Iran) of a clitoral procedure as an intermediate step to eliminating FGA altogether (Latham 2016). The same author cites work in Somalia that has had some success in modifying the severity of FGA procedures, albeit without eliminating them. However, the experiences described above suggest that harm reduction strategies in FGA would be effective, as they have been in other contexts.

Substitution of nicotine patches or gum often weans people off carcinogenic tobacco products (Cahn and Siegel 2011), though some studies show only a limited effect (Bullen et al. 2013). Needle exchange programs have been regarded as one of the greatest successes of the harm reduction approach (Drucker et al. 1998). Here, most of the harm reduction is for those not yet infected; self protection may be less of a motive for someone already infected. However, someone infected with one disease transmitted through fluid exchange may still be infected by a second serious disease; someone who is HIV positive may acquire hepatitis C, for example. So there remains an element of self interest in compliance with needle exchange programs.

Arguably, these examples are not comparable to use of harm reduction in ritual genital procedures. The change in behavior among tobacco users reduces harm for the user herself, while the measures we describe regarding VN and MBP reduce harm for someone other than the actor—namely, for the actor's child. For example, healthcare workers are poorly compliant with voluntary programs of infection reduction such as influenza immunization (Wicker et al. 2009) and hand cleansing between patients (Pittet et al. 1999). Arguably, these examples are not entirely apt either. The beneficiaries are patients, while the beneficiaries of harm reduction measures directed toward ritual genital procedures are the actors' own children, who presumably are more important to the actors involved than patients are to healthcare workers. Even if only 20–40% of parents accepted the harm reduction strategies proposed for FGA and MBP (comparable to those compliant with immunization and hand hygiene), many children would benefit. Perhaps the closest analog to harm reduction programs for the two procedures we discuss is maternal immunization to protect against diseases that are trivial in adults but serious in offspring. When the rubella vaccine was first introduced, many women of childbearing age were not immune, and were susceptible to the teratogenic effects of infection if they contracted rubella during pregnancy. Vaccination causes transient aching and malaise in many recipients, and rare serious complications. Nevertheless, most non-immune hospital workers accepted a proffered rubella immunization (Weiss et al. 1979). Maternal immunization to prevent intrauterine rubella infection seems to be closely analogous to measures proposed to protect infants against neonatal herpes transmission through MBP; indeed, such immunization involves even more effort by the mother. Furthermore, most parents would subject their children to mild discomfort to avoid serious illness or injury (e.g. though immunization or through immobilization in car seats); this is analogous to subjecting a girl to VN to avoid a very painful and deforming FGA procedure. Both respect for autonomy and beneficial considerations justify harm reduction programs (Ruderman 2013). While parents should act in the interests of their children, autonomy allows for and protects the ability of a parent to involve children in religious and cultural rites absent severe foreseeable risks. Blanket condemnation of FGA and circumcision is paternalistic, but providing a safer avenue for expression of religious and cultural beliefs demonstrates respect. While it is outside of the scope of the current discussion to review the ethics of either MBP or VN and we assume for sake of this discussion that both are unethical, harm reduction programs rather than punishment ought to be the societal response given the negative consequences associated with suppressing MBP and the VN.

Applying Theory: Judges and Doctors

Legislators do not punish people who violate the laws they enact; others must carry out their wishes. We are aware of only one case in a common law court involving VN, an agency application to terminate parental rights on Muslim siblings 3 and 4 years old because of an alleged VN on one child (*Matter of B & G (*Matter of B and G (Children) (No 2) Leeds City Council and M, F, B and G, [2015] EWFC 3, Case No: LJ13C00295, Family Court, Leeds, England, 2015)). Judge Sir James Munby ruled that physical evidence did not demonstrate FGA (¶¶47–51). He concluded that future cutting of the girl's genitalia was unlikely in any case (¶52). He observed that nicking was "much less invasive than male circumcision" (¶60). Therefore, a finding of FGA should not necessarily lead to termination of care proceedings since FGA generally is performed only once (¶¶76–77) and, implicitly, because the effects of procedures such as a VN are minimal.

Judge Munby's opinion was a masterpiece of wisdom. He concluded that whatever had happened was no more than minimal, and that further damage was improbable. He implied that termination of parental care would do more harm than good. The judge, however, expressed this as *obiter dictum*, since he had ruled there had been no injury and needed to say no more about FGA.[17] In essence, Judge Munby

[17] A determination of fact ordinarily is not subject to reversal on appeal in a common law court. Appellate courts can only reverse errors in application of law. By failing to find that FGA had been performed, the judge insulated his judgment from hostile appellate review.

declined to break up a family on the basis of a vulvar nick. The facts that the case was even litigated shows how laws punishing minor offenses can lead to abuse of authority by government agents, and how even distinguished jurists may feel obliged to find stratagems to avoid enforcing such laws.

The legality of a medical procedure does not imply that it is ethical, though. Furthermore, health care professionals may have moral scruples outside normative medical ethics. We suggest that a doctor who contemplates performing VN should do so only if all custodial parents consent. If the child upon whom it is to be performed understands the procedure and is able to give or withhold assent, her views should be given strong consideration, depending on her age, degree of understanding, and the overall situation (e.g., whether refusal would lead to performance of a more extensive procedure). A doctor should not be faulted for refusing to perform VN in jurisdictions in which is illegal, though; we believe that respect for law is a *prima facie* principle. Physicians should also decline to perform VN if they have moral scruples which preclude their participation. Physicians would not ordinarily perform MBP, but under some circumstances might be required or permitted to report suspicion that MBP had been performed. The doctor's ethical obligation to report possible abuse or reportable infections may be highly case-specific. The principles that must be balanced include the helpful and harmful effects of alternative actions, the consequences to all parties (including the doctor herself) of disobeying the law, the degree to which the parties involved took precautions to assure the health of the children involved, the ethical obligation to obey the law, and the legitimacy of the legal system. It goes without saying that physicians should not perform procedures that they consider to be immoral or against the patient's interest. A physician has a fiduciary responsibility toward any child for which she has clinical responsibility. This requires that the physician do that which is best for the child under the total circumstances, and these best interests include not just physical health, but the child's social, psychological, and spiritual well-being, and considerations of the likely alternatives to the physician's action and inaction.

This is not to say that there are no minority practices that should be forbidden. We oppose permitting parents to withhold medical care on religious (or other) grounds for serious illness. Here, the magnitude of the harm to the child is so great that the practice fails to meet reasonable criteria for toleration (Levin et al. 2016). However, all faith healing should be treated identically. Members of Christian Science, which recommends treating almost all illnesses via prayer and eschewing medical treatment (Offit 2015), are largely ethnically western European and socioeconomically privileged.[18] However, Christian Science faith healing should be no more or less privileged than faith healing by a Hmong shaman (Fadiman 1997).

[18] Its present and past adherents include two former CIA directors (Stansfield Turner and William Webster), high Nixon aides John Ehrlichman and H.R. Haldeman, and former Treasury Secretary and Goldman Sachs CEO Henry Paulson.

Conclusion

Harm reduction, education, and anticipation of assimilation, rather than punishment, are appropriate tools to combat practices such as VN and MBP that society disapproves of but for which punishment is likely to do more harm than good. Governments should ordinarily not go beyond these educational and harm reduction measures. Our reasons can be encapsulated as follows: Even if FGA and MBP are acts that should not be performed, utilitarian-based punishment, if effective, would be so severe as to do more harm than good. Any significant retributional punishment would be disproportionate to the offense caused by VN or MBP. Finally, expressive punishment would be illiberal. Indeed, when expressive punishment is directed at activities closely associated with clearly identified minority groups, it approaches explicit bigotry. Consequently, a strategy of punishment for VN or MBP would do more harm than good and would, itself, run counter to the principles of a liberal society.

Tools remaining for government use include harm reduction strategies and education. Those who believe that governments should act robustly to protect the health and safety of citizens of the state likely will have no argument in principle with harm reduction and educational programs, though the appropriateness of a particular program is dependent on the facts of the situation.

VN and MBP are exemplars of other practices, some of which may reasonably be punished, and others not. For example, punitive strategies to compel polio vaccination for children may be more justifiable than punitive strategies to compel a hypothetical vaccination against the common cold. Those advocating or considering possible punitive public health measures should consider the utilitarian and retributive ramifications, and should be aware that merely expressive punishment may deviate from liberal principles. In this manner, pursuit of a perfectly ordered society becomes the enemy of a good society.

Acknowledgment Dr. Arora is funded by the Clinical and Translational Science Collaborative of Cleveland, KL2TR000440 from the National Center for Advancing Translational Sciences (NCATS) component of the National Institutes of Health and NIH roadmap for Medical Research. This manuscript is solely the responsibility of the authors and do not necessarily represent the official views of the NIH.

The authors have no conflicts of interest to declare.

References

American Academy of Pediatrics (2010a) Ritual genital cutting of female minors. Pediatrics 125(5):1088–1093

American Academy of Pediatrics (2010b) AAP publications reaffirmed and retired. Pediatrics 126(1):177

Arora KS, Jacobs AJ (2016) Female genital alteration: a compromise solution. J Med Ethics 42(3):148–154

Bengoetxea J (2013) Ultima ratio and the judicial application of law. Oñati Socio-Legal Series 3(1):107–124

Blocker JS Jr (2006) Did prohibition really work? Alcohol prohibition as a public health innovation. Am J Public Health 96(2):233–243

Borland J (1994) #184 Sentence Enhancement. Repeat Offenders. California Voter Foundation. http://www.calvoter.org/archive/94general/props/184.html. Accessed 12 Apr 2015

Broyde MJ (2002) Informing on others for violating American law: a Jewish law view. J Halacha Contemp Soc 43:5

Bullen C, Howe C, Laugesen M, McRobbie H, Parag V, Williman J, Walker N (2013) Electronic cigarettes for smoking cessation: a randomised controlled trial. Lancet 382(9905):1629–1637

Bushman, RL (1960, Autumn) Mormon Persecution in Missouri, 1833. Brigham Young University Studies, III, 11–20

Cahn Z, Siegel M (2011) Electronic cigarettes as a harm reduction strategy for tobacco control: a step forward or a repeat of past mistakes. J Public Health Policy 32(1):16–31

Centers for Disease Control and Prevention (CDC) (2012) Neonatal herpes simplex virus infection following Jewish ritual circumcisions that included direct orogenital suction—New York City, 2000–2011. MMWR Morb Mortal Wkly Rep 61(22):405

Centers for Disease Control and Prevention (2015). Trends in current cigarette smoking among high school students and adults, United States, 1965–2011. http://www.cdc.gov/tobacco/data_statistics/tables/trends/cig_smoking/. Accessed 12 Apr 2015

Cherpes TL, Matthews DB, Maryak SA (2012) Neonatal herpes simplex virus infection. Clin Obstet Gynecol 55(4):938

Coleman DL (1998) The Seattle compromise: multicultural sensitivity and Americanization. Duke Law J 47(4):717–783

Corbett, S (20 Jan 2008) A cutting tradition. New York Times Magazine. http://www.nytimes.com/2008/01/20/magazine/20circumcision-t.html. Accessed 2 Dec 2015

Drucker E, Lurie P, Wodak A, Alcabes P (1998) Measuring harm reduction: the effects of needle and syringe exchange programs and methadone maintenance on the ecology of HIV. AIDS 12(Suppl A):S217–S230

EIGE (2013) European Institute for Gender Equality Report: female genital mutilation in the European Union and Croatia. http://eige.europa.eu/sites/default/files/documents/eige-report-fgm-in-the-eu-and-croatia.pdf. Accessed 14 Feb 2016

Eke N, Nkanginieme KEO (2006) Female genital mutilation and obstetric outcome. Lancet 367(9525):1799–1800

Fadiman A (1997) The spirit catches you and you fall down: a Hmong child, her American Doctors, and the collision of two cultures. Farrar, Straus & Giroux, New York

Forslind E (15 Jan 2015) 38 000 Kvinnor och Flickor i Sverige Kan Vara Könsstympade. Vårdfokustidning för Vårdförbundet (in Swedish). https://www.vardforbundet.se/Vardfokus/Webbnyheter/2015/Januari/38-000-kvinnor-och-flickor-i-Sverige-kan-vara-omskurna/. Accessed 20 Feb 2015

Galeotti AE (2007) Relativism, universalism, and applied ethics: the case of female circumcision. Constellations 14(1):91–111

Gallup Organization (2015) Alcohol and drinking. http://www.gallup.com/poll/1582/alcohol-drinking.aspx. Accessed 13 Apr 2015

Gillespie D, Melching M (2010) The transformative power of democracy and human rights in nonformal education: the case of Tostan. Adult Educ Q 60(5):477–498

Goldberg D (9 Sept 2015) Board of Health Repeals Metzitzah B'peh Consent Form. Politico New York http://www.capitalnewyork.com/article/city-hall/2015/09/8576326/board-health-repeals-metzitzah-bpeh-consent-form. Accessed 7 Oct 2015

Gruenbaum E (2005) Socio-cultural dynamics of female genital cutting: research findings, gaps, and directions. Cult Health Sex 7(5):429–441

Habermas J (1988) Between facts and norms: contributions to a discourse theory of law and democracy (trans. W. Rehg). MIT Press, Cambridge

Harris A, Evans H, Beckett K (2011) Courtesy stigma and monetary sanctions toward a socio-cultural theory of punishment. Am Sociol Rev 76(2):234–264
Huo YJ, Smith HJ, Tyler TR, Allan Lind E (1996) Superordinate identification, subgroup identification, and justice concerns: is separatism the problem; is assimilation the answer? Psychol Sci 7(1):40–45
Jackson J, Bradford B, Hough M, Myhill A, Quinton P, Tyler TR (2012) Why do people comply with the law? Legitimacy and the influence of legal institutions. Br J Criminol 52(6):1051–1071
Jareborg N (2004) Criminalization as last resort (Ultima Ratio). Ohio State J Crim Law 2(2):521–534
Kerlee JC (2004) Too much religious freedom? Infants infected with herpes after Jewish mohel applies oral suction to circumcised penises. J Law Health 19(2):297–411
KNMG (Koninklijke Nederlandsche Maatschappij tot bevordering der Geneeskunst; Royal Dutch Medical Association) (2010) Nontherapeutic circumcision of male minors. http://knmg.artsennet.nl/publicatie/77942/nontherapeutic-circumcisionof-male-minors-2010.htm
Korobkin ND (28 Dec 2006). Metzitzah B'peh controversy: rabbinic polemics and applying the lessons of history. Jewish Action. https://www.ou.org/jewish_action/12/2006/metzitzah_bpeh_rabbinic_polemics/. Accessed 14 Feb 2016
Latham S (2016) The campaign against female genital cutting: empowering women or reinforcing global inequity? Ethics Soc Welf 10(1):1–14
Levin HY, Jacobs Allan J, Arora Kavita S (2016) To accommodate or not to accommodate: (When) should the state regulate religion to protect the rights of children and third parties? Washington & Lee Law Review 73:915–1016
Leye E, Powell RA, Nienhuis G, Claeys P, Temmerman M (2006) Health care in Europe for women with genital mutilation. Health Care Women Int 27(4):362–378
Leye E, Deblonde J, García-Añón J, Johnsdotter S, Kwateng-Kluvitse A, Weil-Curiel L, Temmerman M (2007) An analysis of the implementation of laws with regard to female genital mutilation in Europe. Crime Law Soc Chang 47(1):1–31
Marlatt GA, Tapert SF (1993) Harm reduction: reducing the risks of addictive behaviors. In: Baer JS, Marlatt GA, McMahon R (eds) Addictive behaviors across the lifespan. Sage, Newbury Park, pp 243–273
Marlatt GA (1996) Harm reduction: come as you are. Addict Behav 21(6):779–788
Marlatt GA, Witkiewitz K (2002) Harm reduction approaches to alcohol use: health promotion, prevention, and treatment. Addict Behav 27(6):867–886
Naser RL, Visher CA (2006) Family members' experiences with incarceration and reentry. West Criminol Rev 7(2):20–31
Obermeyer CM (1999) Female genital surgeries: the known, the unknown, and the unknowable. Med Anthropol Q 13(1):79–106
Obiora LA (1996) Bridges and barricades: rethinking polemics and intransigence in the campaign against female circumcision. Case West Res Law Rev 47(3):275–378
Offit PA (2015) Bad faith: when religious belief undermines modern medicine. Basic Books, New York
One Hundred Professors of Obstetrics and Gynecology (2013) A statement on abortion by 100 professors of obstetrics: 100 years later. Am J Obstet Gynecol 209(3):193–199
Orsi RA (2003) Is the study of lived religion irrelevant to the world we live in? Special presidential plenary address, Society for the Scientific Study of Religion, Salt Lake City, 2 Nov 2002. J Sci Study Relig 42(2):169–174
Otterman S (12 Sept 2012) Denouncing city's move to regulate circumcision. New York Times. http://www.nytimes.com/2012/09/13/nyregion/regulation-of-circumcision-method-divides-some-jews-in-new-york.html. Accessed 14 Feb 2016
Petersilia J (2001) Prisoner reentry: public safety and reintegration challenges. Prison J 81(3): 360–375
Pittet D, Mourouga P, Perneger TV (1999) Compliance with handwashing in a teaching hospital. Ann Intern Med 130(2):126–130

Rawls J (1955) Two concepts of rules. Philos Rev 64(1):3–32
Resnicoff SH (2012) Jewish law and the tragedy of sexual abuse of children: the dilemma within the orthodox Jewish community. Rutgers J Law Relig 13(2):281–362
Ruderman R (2013) Female circumcision: the ethics of harm reduction policies. Michigan J Public Aff 10(1):95–107
Schrad ML (2007) Constitutional blemishes: American alcohol prohibition and repeal as policy punctuation. Policy Stud J 35(3):437–463
Shell-Duncan B (2001) The medicalization of female "circumcision": harm reduction or promotion of a dangerous practice? Soc Sci Med 52(7):1013–1028
Shell-Duncan B (2008) From health to human rights: female genital cutting and the politics of intervention. Am Anthropol 110(2):225–236
Shell-Duncan B, Wander K, Hernlund Y, Moreau A (2013) Legislating change? Responses to criminalizing female genital cutting in Senegal. Law Soc Rev 47(4):803–835
Staub E (1999) The roots of evil: social conditions, culture, personality, and basic human needs. Personal Soc Psychol Rev 3(3):179–192
Steinberg MB, Delnevo CD (2013) Increasing the "smoking age": the right thing to do. Ann Intern Med 159(8):558–559
Tadros V (2011) The ends of harm. Oxford University Press, Oxford
Topping A (5 Feb 2015) FGM affects three times more people in the US than previously thought. The Guardian. http://www.theguardian.com/society/2015/feb/05/fgm-numbers-affected-us-women. Accessed 14 Feb 2016
Tuori KH (2013) Ultima ratio as a constitutional principle. Oñati Socio-Legal Series 3(1):6–20
Tyler TR (2006) Psychological perspectives on legitimacy and legitimation. Annu Rev Psychol 57:375–400
Walzer M (1983) Spheres of justice: a defense of pluralism and equality. Basic Books, New York
Weiss KE, Falvo CE, Buimovici-Klein E, Magill JW, Cooper LZ (1979) Evaluation of an employee health service as a setting for a rubella screening and immunization program. Am J Public Health 69(3):281–283
Wicker S, Rabenau HF, Doerr HW, Allwinn R (2009) Influenza vaccination compliance among health care workers in a German university hospital. Infection 37(3):197–202
World Health Organization (2014) Female genital mutilation and other harmful practices: prevalence of FGM. http://www.who.int/reproductivehealth/topics/fgm/prevalence/en/. Accessed 25 Oct 2014
Wringe B (2012) Collective agents and communicative theories of punishment. J Soc Philos 43(4):436–456
Zaibert L (2013) The instruments of abolition, or why retributivism is the only real justification of punishment. Law Philos 32(1):33–58
Zhu S-H, Lee M, Zhuang Y-L, Gamst A, Wolfson T (2012) Interventions to increase smoking cessation at the population level: how much progress has been made in the last two decades? Tob Control 21(2):110–118

Index

L. Campo-Engelstein, P. Burcher (eds.), *Reproductive Ethics*,
DOI 10.1007/978-3-319-52630-0

MIX
Papier aus verantwortungsvollen Quellen
Paper from responsible sources
FSC® C105338

If you have any concerns about our products,
you can contact us on
ProductSafety@springernature.com

In case Publisher is established outside the EU,
the EU authorized representative is:
Springer Nature Customer Service Center GmbH
Europaplatz 3, 69115 Heidelberg, Germany

Printed by Libri Plureos GmbH
in Hamburg, Germany